ANATOMY 360°

The Ultimate Visual Guide to the Human Body

THUNDER BAY
P·R·E·S·S

San Diego, California

Thunder Bay Press
An imprint of Printers Row Publishing Group
10350 Barnes Canyon Road, Suite 100, San Diego, CA 92121
www.thunderbaybooks.com

Primal Pictures Ltd
4th Floor, Tennyson House
159-165 Great Portland Street
London W1W 5PA, UK

Thunder Bay Press
Publisher: Peter Norton
Associate Publisher: Ana Parker
Publishing/Editorial Team: April Farr, Kelly Larsen, Kathryn C. Dalby
Editorial Team: JoAnn Padgett, Melinda Allman
Production Team: Jonathan Lopes, Rusty von Dyl

Anatomical Images
Primal Pictures Ltd, London, UK, www.primalpictures.com
Editor
Lorna Wilson, Senior Anatomist, Primal Pictures, UK
Author
Dr. Jamie Roebuck, BSc (Hons), MBChB, FHEA, Clinical Education Fellow, Warwick University, UK
Design
Rebecca Painter; Carolina Stenstrom
Cover Design
Nick Harris, www.splinter-group.net; Rusty von Dyl

ISBN: 978-1-68412-280-6

The Library of Congress has cataloged the original Thunder Bay hardcover as follows:

Library of Congress Cataloging-in-Publication Data

Roebuck, Jamie, author.
 Anatomy 360° / by Jamie Roebuck. -- Second edition.
    p. ; cm.
 Anatomy three-hundred and sixty degree
 ISBN 978-1-62686-474-0 (hardcover)
 I. Title. II. Title: Anatomy three-hundred and sixty degree.
 [DNLM: 1. Anatomy--Atlases. QS 17]
 QM25
 611.00222--dc23
                    2015001387

Printed in China
22 21 20 19 18 3 4 5 6 7

ANATOMY
360°

The Ultimate Visual Guide
to the Human Body

CONTENTS

The amazing images found within this book were taken from the interactive products of Primal Pictures. They were generated from a comprehensive 3-D model that took twenty years to complete and are made up of more than 7,000 individual anatomical structures. Any combination of structures can be seen from any angle, allowing the vast array of images you can see in this book to be created.

THE MODEL

Creating the model began with building the skeleton. The bones of a full adult skeleton were scanned using a CT machine and reconstructed as a detailed 3-D computer model. The skeleton

could then act as a framework for the soft tissues of the body that were modeled.

Structures such as the muscles, vessels, nerves, and organs were constructed in a process using data from Primal MRI data and the Visible Human Project.

Visible Human Project

The **Visible Human Project** was created by the U.S. National Library of Medicine and involved the production of a male and female set of cross sectional photographs. The bodies were frozen in a hardening agent then sectioned *axially* (horizontally). Each 0.3mm section was ground away at a time, and the remains photographed. The pixels of the axial data set were reconfigured to produce matching *coronal* and *sagittal* slices.

Segmentation

Our anatomy team viewed the cross-sectional images and used an in-house program to digitally outline every individual anatomical structure at each level, and often in all three planes. This process is known as **segmentation** and took the team over four years to complete. Once each component was labeled throughout the entire data set, the labels were transferred into the 3-D computer graphics package, **Houdini**.

This image shows a label around the **caudate nucleus** of the brain on an axial slice.

Splines

The sets of rings taken were then skinned and manually manipulated to provide smooth realistic shapes.

In Houdini, the graphics team used the labels to create a 3-D shape of each anatomical component. Here are the rings for the lateral ventricles of the brain seen in axial, sagittal, and coronal directions, and combined together.

Final model

Smaller components that could not be segmented, such as small ligaments and vessels were hand-modeled and the 3-D model was textured and rendered to produce the images you can see in this book.

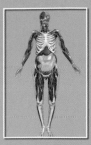

The human body is truly amazing. At this very moment numerous chemical reactions are taking place within the trillions of specialized cells that make up our tissues, organs, and systems. Working together, they allow us to move, to grow, to breathe, to eat, to react to our environment, and even to interact with one another. Thanks to our incredible bodies, we are living organisms that not only survive, but thrive, here on planet Earth. As we take this journey through the human body we will see the remarkable way in which the body is adapted and organized to allow this to happen.

Human anatomy is the branch of science that studies the structure of the human body. In the past, this has involved dissecting human bodies to separate and reveal the different parts. However, new technology such as magnetic resonance imaging (MRI) now allows us to view the structure of the entire human body without making a single cut.

Human anatomy is closely linked with another branch of science called physiology. This is the study of the way in which the body works. It is helpful to keep in mind that the structure of a body part is related to the task that it performs—form (anatomy) is related to function (physiology).

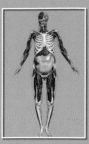

A cell is the smallest individual building block of a living organism. There are many different types of cell, depending upon the substances they produce and the functions they perform. Examples include skin, muscle, and nerve cells. Each cell contains organelles which are specialized regions within a cell that work together to keep the cell alive, and to help it carry out its function.

Did you know?

It is estimated that there are approximately 100,000 billion cells in the average adult human (that's 10 followed by thirteen zeros!)

Cytoplasm
is the part of the cell lying between the plasma membrane and the nucleus. The fluid in the cytoplasm in which the organelles are suspended is known as the cytosol.

Nucleus

Endoplasmic reticulum

Golgi complex

Centrioles

Lysosomes
are organelles that contain substances that can break down waste material and debris within the cell.

Mitochondria

Cytoskeletons
provide structural support for the cell and help move things from one part of the cell to another.

Plasma membrane
is a barrier that separates the inside of the cell from the environment.

Organelles

Nucleus
is the control center of each cell. The nucleus contains most of the cell's genetic information.

Golgi complex
can be thought of as an "intracellular post office." Products made by the cell are labeled and packaged for delivery to other areas, or for export out of the cell.

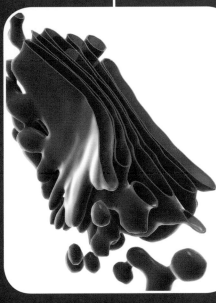

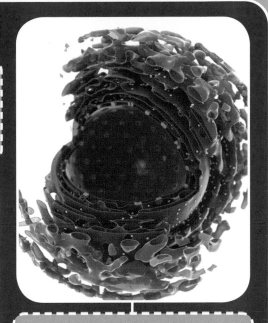

Endoplasmic reticulum
is involved in producing and processing proteins. In some cells it also has a role in the processing of fats and sugars.

Cytoskeleton
provides structural support for the cell as well as helping move things from one part of the cell to another.

Mitochondria
are the "power stations" of the cell. They provide energy using the oxygen and nutrients delivered to the cell.

Centrioles
are specialized structures involved in cell division.

CELL DIVISION

Most cells in the human body are able to divide. There are two methods of cell division, mitosis and meiosis. These processes are divided into different phases.

Body (somatic) cells divide by mitosis. This produces two genetically identical copies of the parent cell, which can replace worn out cells, and allow growth to occur.

Reproductive (germ) cells divide by meiosis. This produces four genetically unique cells called gametes (either sperm or egg cells), which have half (1n) the genetic material of the parent cell. Gametes allow reproduction and the creation of a new human being.

Mitosis

is the type of cell division that takes place in body (somatic) cells.

Prophase
is where the genetic material in the cell nucleus bunches together to form paired X-shaped structures called chromosomes.

Metaphase
is where the chromosomes all line up.

Anaphase
is where the chromosomes divide, and each identical half is pulled to opposite sides of the cell nucleus.

Meiosis

is the type of cell division that takes place in reproductive (germ) cells. It is a two-step process. The key difference to mitosis is that the resulting cells have only half the normal content of genetic material.

Prophase I

Metaphase I

Anaphase I

Telophase I

Prophase
It differs to mitosis in that there is exchange of genetic material between the paired chromosomes. This means that the gametes are genetically different from each other.

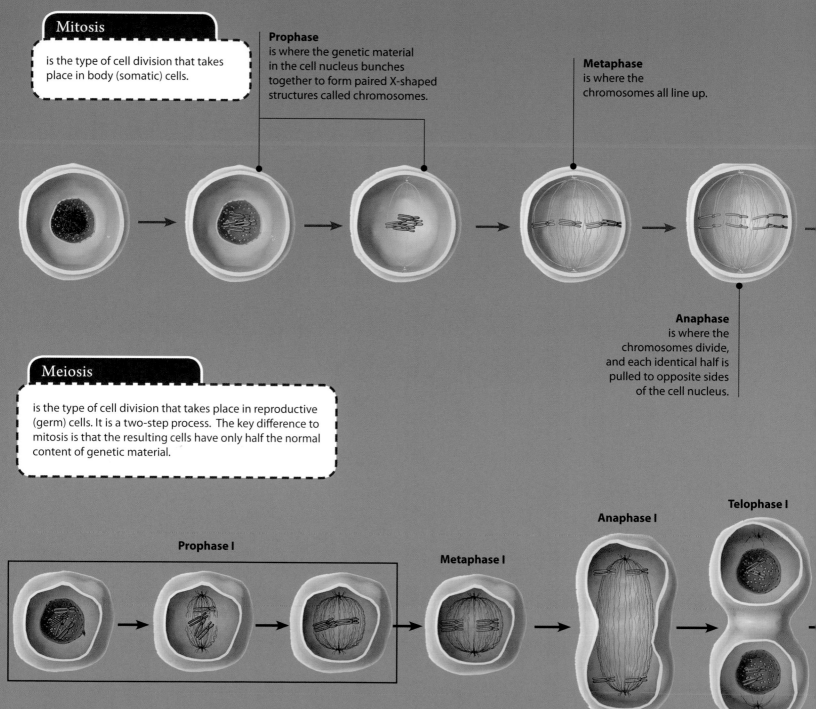

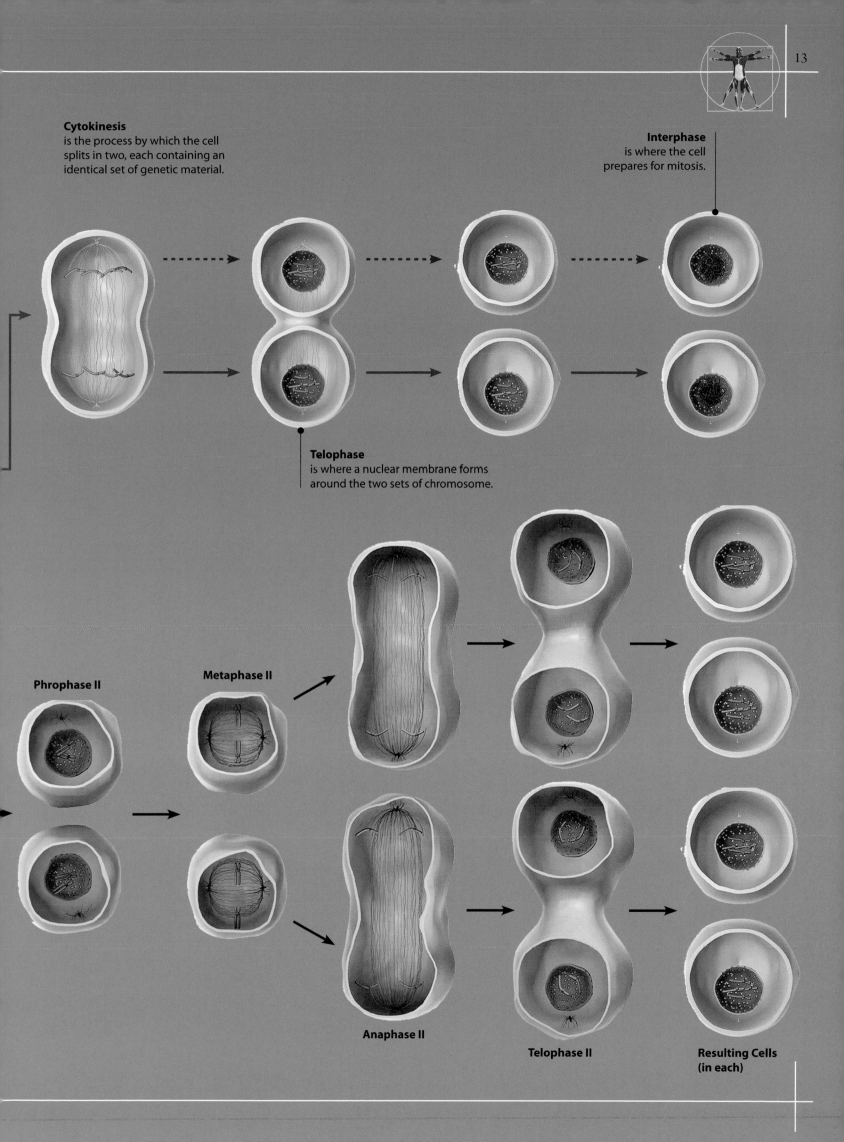

Cytokinesis
is the process by which the cell splits in two, each containing an identical set of genetic material.

Interphase
is where the cell prepares for mitosis.

Telophase
is where a nuclear membrane forms around the two sets of chromosome.

Phrophase II

Metaphase II

Anaphase II

Telophase II

**Resulting Cells
(in each)**

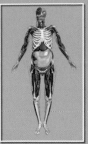

The plasma membrane is a flexible barrier that separates and protects the cell contents from the outside environment. It is selectively permeable, which means that only certain substances can pass through it. This allows the plasma membrane to control what enters and leaves the cell, and to precisely regulate the environment within the cell. The plasma membrane is formed by two layers of phospholipids, within which cholesterol and various specialized proteins are embedded. These proteins are the cell's main way of interacting with the external environment, and they include channels, signals, receptors, carriers, and enzymes.

Glycolipids
are a type of fat (lipid) with a chain of sugars attached. They allow cells to be recognized by other cells.

Cell adhesion molecules
allow cells to bind to one another and to the surrounding connective tissue.

Enzymes
speed up certain chemical reactions either inside or outside the cell.

Cholesterol molecules
help stabilize the plasma membrane.

Phospholipid bilayer
provides a flexible semi-permeable barrier in which various proteins and glycolipids are embedded.

Cell identification molecules allow cells to signal and communicate with each other.

Receptor proteins detect when specific molecules bind to them, often leading to a change in cell behavior.

Carrier proteins can undergo shape changes which allow them to move large molecules from outside the cell to inside the cell.

Channel proteins allow the passage of certain chemical substances, depending upon their size and electrical charge.

ORGANIZATION OF TISSUES

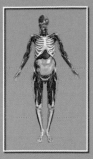

The human body contains trillions of individual cells. To maximize efficiency and function, this vast number of cells are organized and grouped according to the function that they perform. This leads to a hierarchy of cells, tissues, organs, systems, and organisms, where each level increases in both structural and functional complexity.

Cells

A cell is the basic building block of life within an organism. Most cells have a specific function; some produce digestive secretions like the cells lining the gut, while others contract like muscle cells for example.

Tissue

Tissue is formed from collections of cells that perform the same function.

Glandular epithelial tissue is formed from individual gastric secretory cells that combine to produce gastric juices in the stomach.

Smooth muscle tissue in the wall of the stomach is formed from individual smooth muscle cells working together to mix up food and move it along the gut.

Organs

Organs are formed by different tissues that work together to perform a function.

The stomach is an organ that is formed from glandular epithelial tissue, connective tissue, nervous tissue and smooth muscle tissue.

Organisms

An organism is made up of many different systems all working together.

Systems

Systems are any organs working together to produce the same effect. Some organs may belong to more than one system.

Digestive system includes the mouth, salivary glands, throat, esophagus, stomach, liver, gallbladder, pancreas, and small and large intestines.

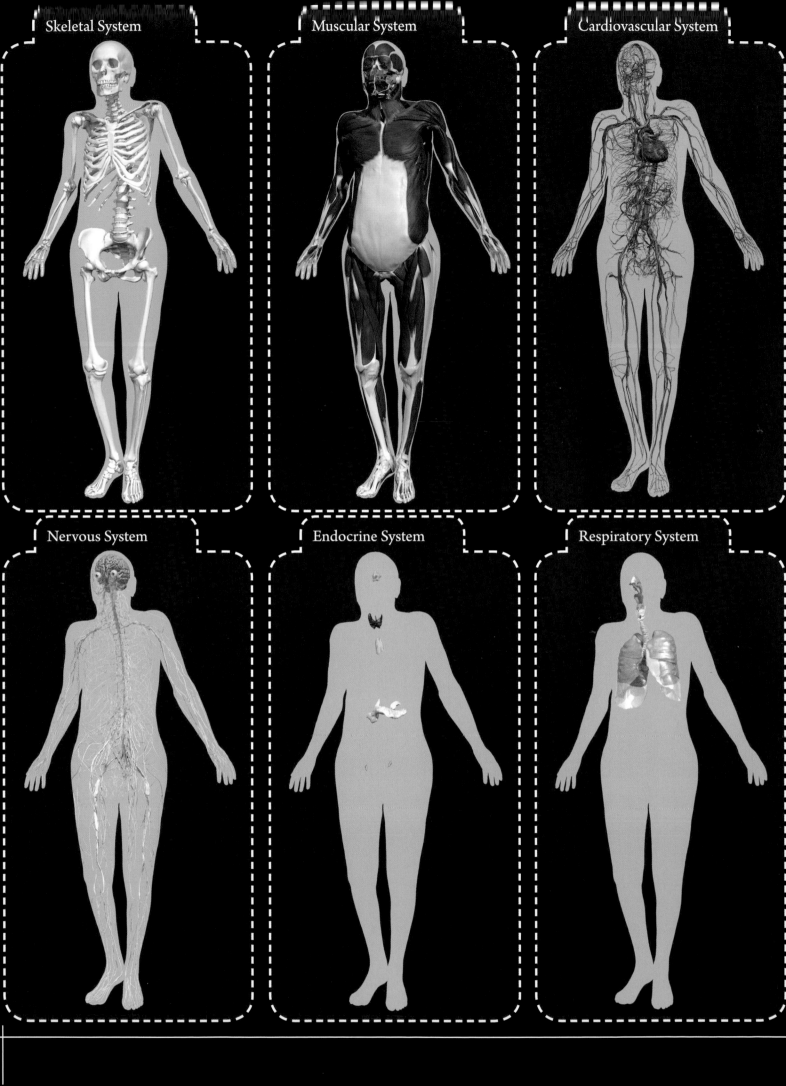

Skeletal System

Muscular System

Cardiovascular System

Nervous System

Endocrine System

Respiratory System

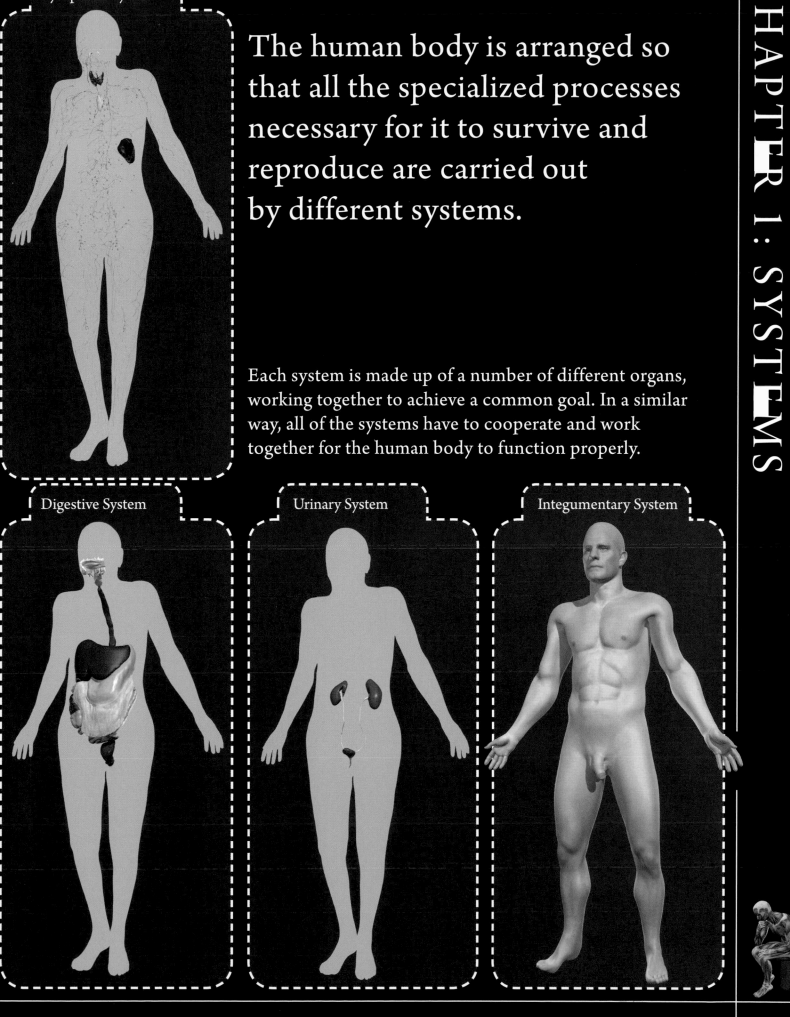

Lymphatic System

The human body is arranged so that all the specialized processes necessary for it to survive and reproduce are carried out by different systems.

Each system is made up of a number of different organs, working together to achieve a common goal. In a similar way, all of the systems have to cooperate and work together for the human body to function properly.

Digestive System

Urinary System

Integumentary System

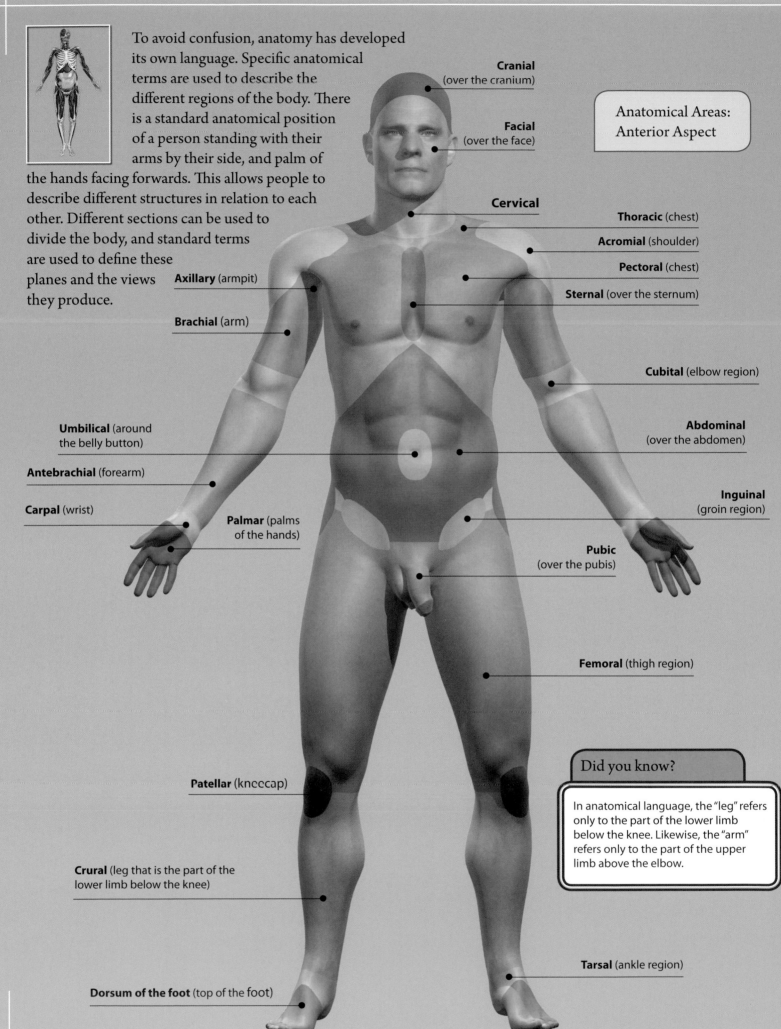

To avoid confusion, anatomy has developed its own language. Specific anatomical terms are used to describe the different regions of the body. There is a standard anatomical position of a person standing with their arms by their side, and palm of the hands facing forwards. This allows people to describe different structures in relation to each other. Different sections can be used to divide the body, and standard terms are used to define these planes and the views they produce.

Anatomical Areas: Anterior Aspect

Cranial (over the cranium)

Facial (over the face)

Cervical

Thoracic (chest)

Acromial (shoulder)

Pectoral (chest)

Sternal (over the sternum)

Axillary (armpit)

Brachial (arm)

Cubital (elbow region)

Umbilical (around the belly button)

Abdominal (over the abdomen)

Antebrachial (forearm)

Inguinal (groin region)

Carpal (wrist)

Palmar (palms of the hands)

Pubic (over the pubis)

Femoral (thigh region)

Patellar (kneecap)

Did you know?

In anatomical language, the "leg" refers only to the part of the lower limb below the knee. Likewise, the "arm" refers only to the part of the upper limb above the elbow.

Crural (leg that is the part of the lower limb below the knee)

Tarsal (ankle region)

Dorsum of the foot (top of the foot)

Anatomical Areas: Posterior Aspect

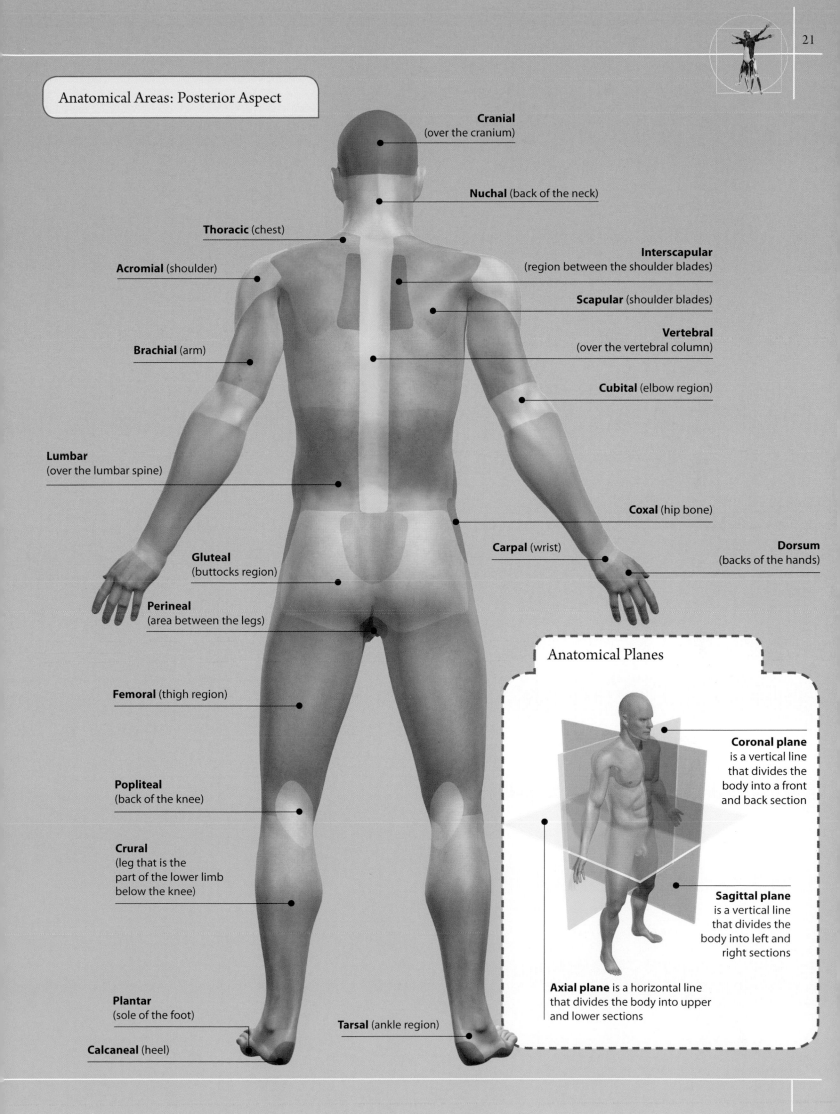

Cranial (over the cranium)

Nuchal (back of the neck)

Thoracic (chest)

Interscapular (region between the shoulder blades)

Acromial (shoulder)

Scapular (shoulder blades)

Vertebral (over the vertebral column)

Brachial (arm)

Cubital (elbow region)

Lumbar (over the lumbar spine)

Coxal (hip bone)

Carpal (wrist)

Dorsum (backs of the hands)

Gluteal (buttocks region)

Perineal (area between the legs)

Femoral (thigh region)

Popliteal (back of the knee)

Crural (leg that is the part of the lower limb below the knee)

Plantar (sole of the foot)

Tarsal (ankle region)

Calcaneal (heel)

Anatomical Planes

Coronal plane is a vertical line that divides the body into a front and back section

Sagittal plane is a vertical line that divides the body into left and right sections

Axial plane is a horizontal line that divides the body into upper and lower sections

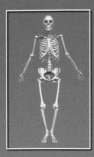

The bones of the skeletal system are living organs that provide protection, support, and movement. They are the body's main store of calcium, and the site of blood cell production. The structure of a typical long bone illustrates the adaptations that make it both strong enough to bear our weight and protect us, yet light enough for us to move it.

Epiphyseal line
is the site of the growth plate in mature long bones.

Trabecular bone or spongy bone
consists of struts of bone. It is found in the epiphyses and lining the inner parts of a long bone.

Anatomy of a Long Bone

Epiphysis
is the end of a long bone.

Metaphysis
is the area between the diaphysis and epiphysis.

Diaphysis
is the shaft of a long bone.

Medullary cavity,
or marrow cavity, is the area where blood cells are produced.

Cortical bone
is the dense outer layer of bone.

Osteons
are repeating layers of bone tissue and bone cells formed around a central canal that contains blood vessels and nerves.

Cortical bone,
or compact bone, is made up of numerous osteons. This makes it very strong and dense.

Periosteum
is a layer of fibrous tissue that lines, protects, and nourishes most of the outer bone surface. It allows ligaments and tendons to attach to a bone.

Trabecular or spongy bone
consists of columns and struts of bone. It provides strength while minimizing weight.

Spongy Bone

Classification of Bone

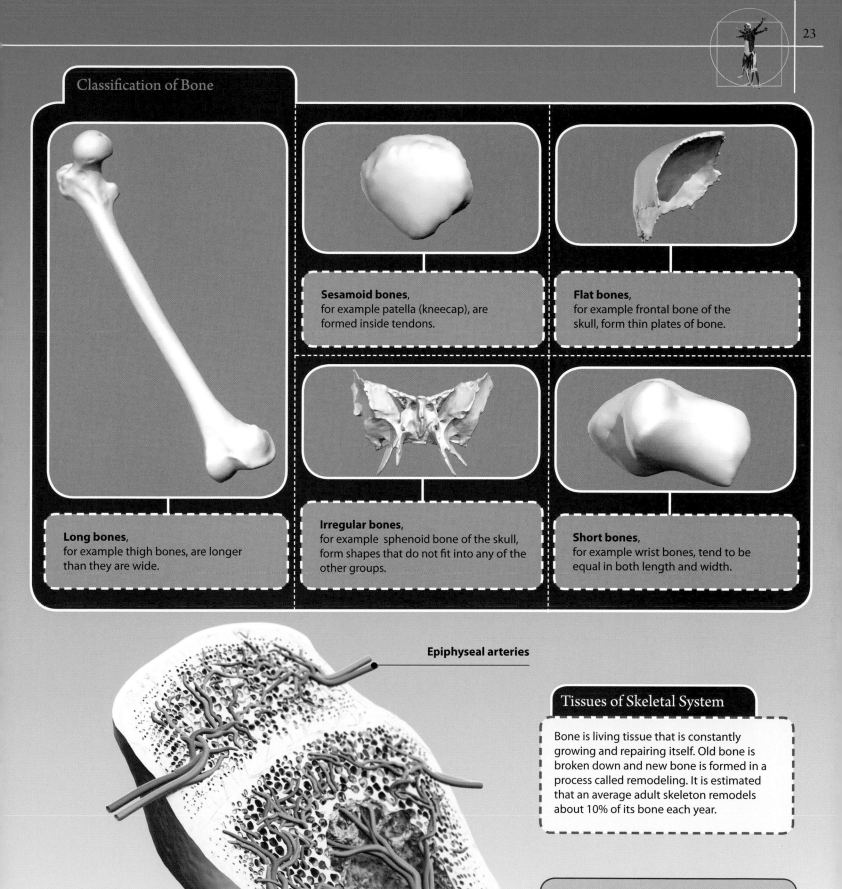

Sesamoid bones,
for example patella (kneecap), are
formed inside tendons.

Flat bones,
for example frontal bone of the
skull, form thin plates of bone.

Long bones,
for example thigh bones, are longer
than they are wide.

Irregular bones,
for example sphenoid bone of the skull,
form shapes that do not fit into any of the
other groups.

Short bones,
for example wrist bones, tend to be
equal in both length and width.

Epiphyseal arteries

Tissues of Skeletal System

Bone is living tissue that is constantly
growing and repairing itself. Old bone is
broken down and new bone is formed in a
process called remodeling. It is estimated
that an average adult skeleton remodels
about 10% of its bone each year.

Blood Supply to Bone

Metaphyseal arteries

Nutrient arteries
supply blood to the medullary
cavity and diaphysis arteries.

The skeletal system is made up of 206 different bones. It provides a strong protective framework for the rest of the body. Joints are formed when two bones meet. There are six main types of joint that allow different ranges of movement. They are named according to the shapes of the bone surfaces that form the joint.

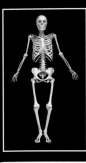

Skeletal System

Skull
protects our brain, as well as forming our face.

Radius
is the shorter bone in the forearm.

Ulna
is the longer bone in the forearm and forms the "point" of our elbow.

Humerus
is the arm bone that links the forearm to the shoulder.

Femur
or thigh bone, is the largest bone in the body.

Patella (or kneecap)

Tibia (or shin bone)

Scapula
or shoulder blade.

Sternum
or breastbone.

Ribs
form a protective cage around the heart and lungs.

Vertebral column
protects the spinal cord and nerves.

Fibula
forms the outer part of the leg and ankle.

Synovial Joints

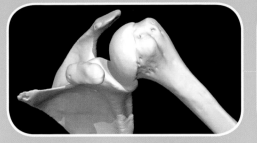

Ball and socket joints
for example shoulder, allow a wide range of movement, as the "ball" end of one bone moves freely within the "socket" of the other bone.

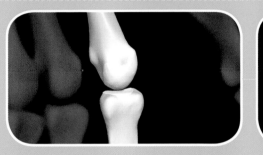
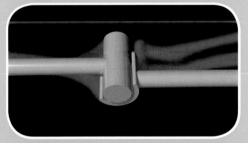

Condyloid joints
for example knuckles, consist of a rounded end of one bone moving within a shallow depression on the other bone.

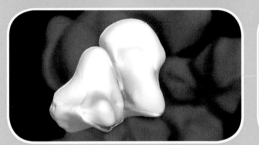
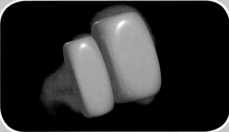

Gliding joints
for example wrist, have two flat bone surfaces moving past each other.

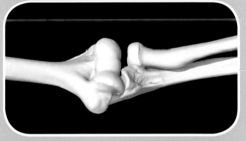
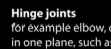

Hinge joints
for example elbow, only allow movement in one plane, such as bending and straightening.

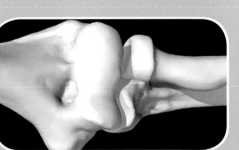

Pivot joints
for example forearm, allow rotational movements to take place.

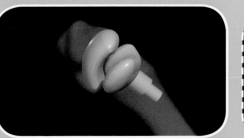

Saddle joint
for example thumb, allows movements in a number of different directions due to the shape of the bone surfaces.

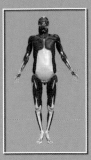

The main role of the muscular system is to produce movement. It does this through specialized muscle cells that are able to contract and alter their length. There are three main types of muscle tissue: skeletal, smooth, and cardiac. Skeletal muscle is attached to the skeletal system via tendons. It is under voluntary control. Smooth and cardiac muscle are termed involuntary muscles, as their contractions can occur without conscious control. Smooth muscle is mainly located within the walls of the internal organs. Cardiac muscle is found in the heart.

Smooth muscle

Smooth muscle helps control the size of internal organ passageways and the movement of substances through them.

Nucleus

Mitochondria produce energy for muscle contraction.

Contractile filaments produce shortening of each smooth muscle cell.

Dense bodies are points of attachment for parts of the contractile filaments.

Smooth muscle cells are spindle shaped and form close connections with surrounding cells.

Myofibrils are organized collections of specialized proteins called myofilaments, that use energy to cause muscle contraction.

Skeletal muscle

Skeletal muscle has a highly organized structure.

Muscle fibers are individual muscle cells. Each fiber is packed with myofibrils that run the entire length of the muscle.

Fascicles are bundles of muscle fibers held together by connective tissue.

Sarcomere

Sarcomeres are regular repeating divisions of the myofibrils. The myofilaments within each sarcomere use energy to slide past each other, causing shortening of the entire muscle (contraction).

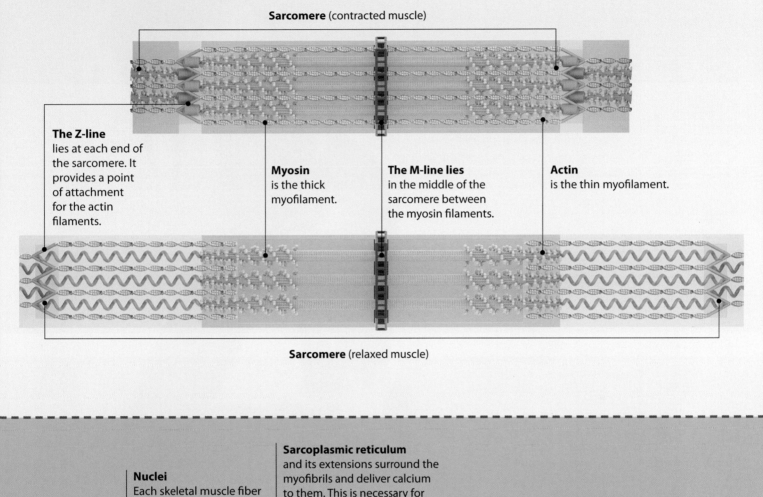

Sarcomere (contracted muscle)

The Z-line
lies at each end of the sarcomere. It provides a point of attachment for the actin filaments.

Myosin
is the thick myofilament.

The M-line lies
in the middle of the sarcomere between the myosin filaments.

Actin
is the thin myofilament.

Sarcomere (relaxed muscle)

Nuclei
Each skeletal muscle fiber contains multiple nuclei.

Sarcoplasmic reticulum
and its extensions surround the myofibrils and deliver calcium to them. This is necessary for contraction to occur.

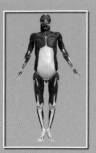

There are over 600 muscles in the human body. Working together they allow us to run and jump, to skip and dance, to eat and speak. Skeletal muscles come in various shapes and sizes, and can be grouped according to the way in which their fibers are organized. Muscles are connected to bones by tough fibrous tissue called tendons.

Latissimus dorsi
pulls the arm toward the body.

Gluteus maximus
forms the buttocks. It is the largest muscle in the body.

Gastrocnemius
is one of the calf muscles at the back of the leg.

Calcaneal tendon,
or Achilles tendon, is the tough fibrous tissue that connects the calf muscles to the heel bone (calcaneum).

Biceps femoris
is one of the hamstring muscles that bend the knee.

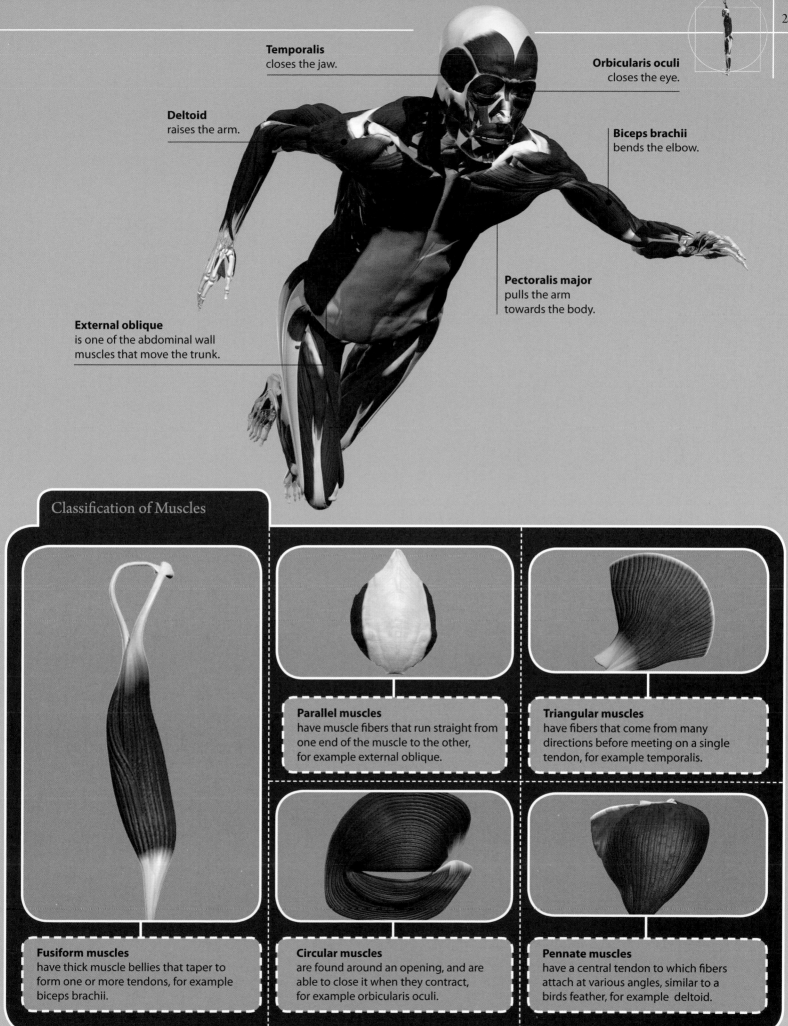

Temporalis
closes the jaw.

Orbicularis oculi
closes the eye.

Deltoid
raises the arm.

Biceps brachii
bends the elbow.

Pectoralis major
pulls the arm
towards the body.

External oblique
is one of the abdominal wall
muscles that move the trunk.

Classification of Muscles

Fusiform muscles
have thick muscle bellies that taper to
form one or more tendons, for example
biceps brachii.

Parallel muscles
have muscle fibers that run straight from
one end of the muscle to the other,
for example external oblique.

Triangular muscles
have fibers that come from many
directions before meeting on a single
tendon, for example temporalis.

Circular muscles
are found around an opening, and are
able to close it when they contract,
for example orbicularis oculi.

Pennate muscles
have a central tendon to which fibers
attach at various angles, similar to a
birds feather, for example deltoid.

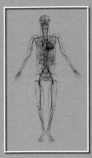

The cardiovascular system is made up of the heart and blood vessels. Together, they provide a constant supply of oxygen and nutrient-rich blood to all the tissues of the body, while removing any waste products. The heart is a muscular organ made up of specialized cardiac muscle tissue. It pumps blood around the body's blood vessels, of which there are three main types: arteries, capillaries, and veins.

Cardiac Muscle

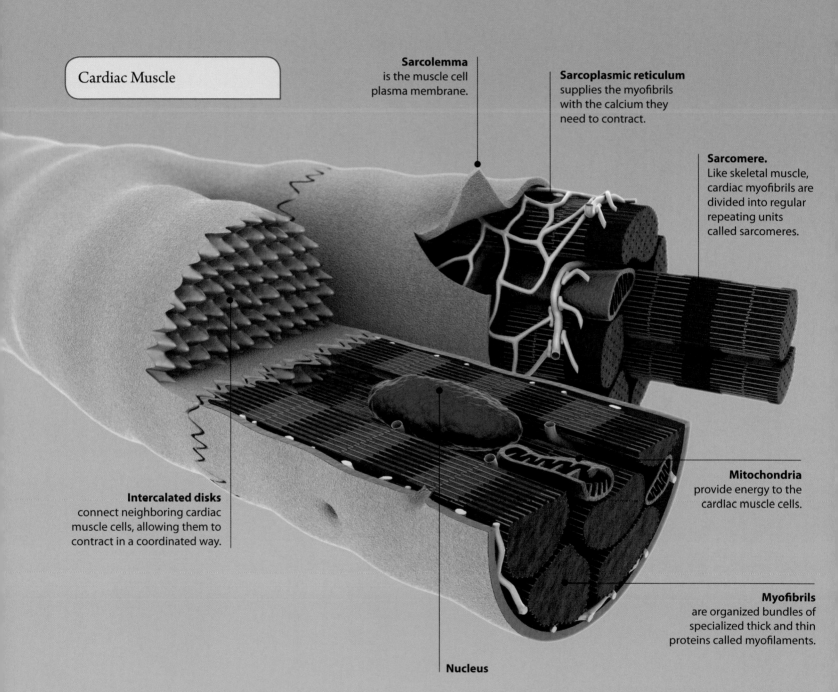

Sarcolemma
is the muscle cell
plasma membrane.

Sarcoplasmic reticulum
supplies the myofibrils
with the calcium they
need to contract.

Sarcomere.
Like skeletal muscle,
cardiac myofibrils are
divided into regular
repeating units
called sarcomeres.

Mitochondria
provide energy to the
cardiac muscle cells.

Intercalated disks
connect neighboring cardiac
muscle cells, allowing them to
contract in a coordinated way.

Myofibrils
are organized bundles of
specialized thick and thin
proteins called myofilaments.

Nucleus

Medium (muscular) artery
for example radial artery

Medium vein
for example renal vein

Valve

Arteries

Arteries carry blood away from the heart to the capillaries. Their walls are relatively thick with muscle and elastic tissue to cope with the high pressure of blood when it leaves the heart.

Veins

Veins return blood from the capillaries back to the heart. As the blood is at relatively low pressure, their walls are thinner than arteries. Some veins have valves, which keep blood flowing in the right direction.

Tunica adventitia
is the outermost layer that attaches the blood vessel to the surrounding tissues.

Tunica intima
is the innermost layer of flattened cells that line the blood vessel.

Large (elastic) artery
for example aorta.

Large vein
for example internal jugular vein.

Tunica media
is the middle layer and contains various amounts of smooth muscle and elastic tissue depending on the blood vessel type.

Capillaries

Capillaries are the smallest blood vessels. Their thin walls allow substances to easily move between the tissues and the blood.

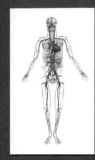

Blood is a living liquid made up of cells and fluid. It is pumped around the blood vessels by the heart. Its main role is to deliver oxygen and nutrients to the tissues of the body, and carry waste products away.

The fluid part of blood is called plasma. It contains proteins, salts, and nutrients.

There are different types of cells in the blood. Red cells carry oxygen to the tissues; white cells fight infection; and platelets help to form clots. They are all produced in the bone marrow.

Erythrocytes (red blood cells)

Lymphocytes

Platelets

Neutrophils

Basophils

Did you know?

A single drop of blood contains approximately 250 million red blood cells.
Platelets are not cells. Instead they are tiny fragments of much larger cells called megakaryocytes.

Monocytes

Eosinphil

Formed Elements of Blood

Erythrocytes (red blood cells)
transport oxygen from the lungs to the tissues. The red color is due to the presence of a specialized oxygen-carrying protein called hemoglobin.

Platelets
are small fragments of cells that help our blood to clot.

Lymphocytes
are white blood cells that can fight a wide range of infectious organisms. There are three types of lymphocyte: B-cells, T-cells, and natural killer (NK) cells.

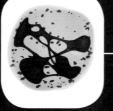

Monocytes
are white blood cells that can move into the tissues, where they are called macrophages. They remove cell debris and infectious organisms by engulfing them—a process known as phagocytosis.

Neutrophils
are the most common white blood cell. They have irregularly shaped nuclei. Their granules contain substances used to kill infectious organisms.

Eosinophil
is a type of white blood cell particularly involved in fighting parasite infections.

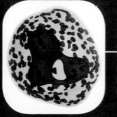

Basophils
are the least abundant white blood cells. Their granules contain substances that cause inflammation when released.

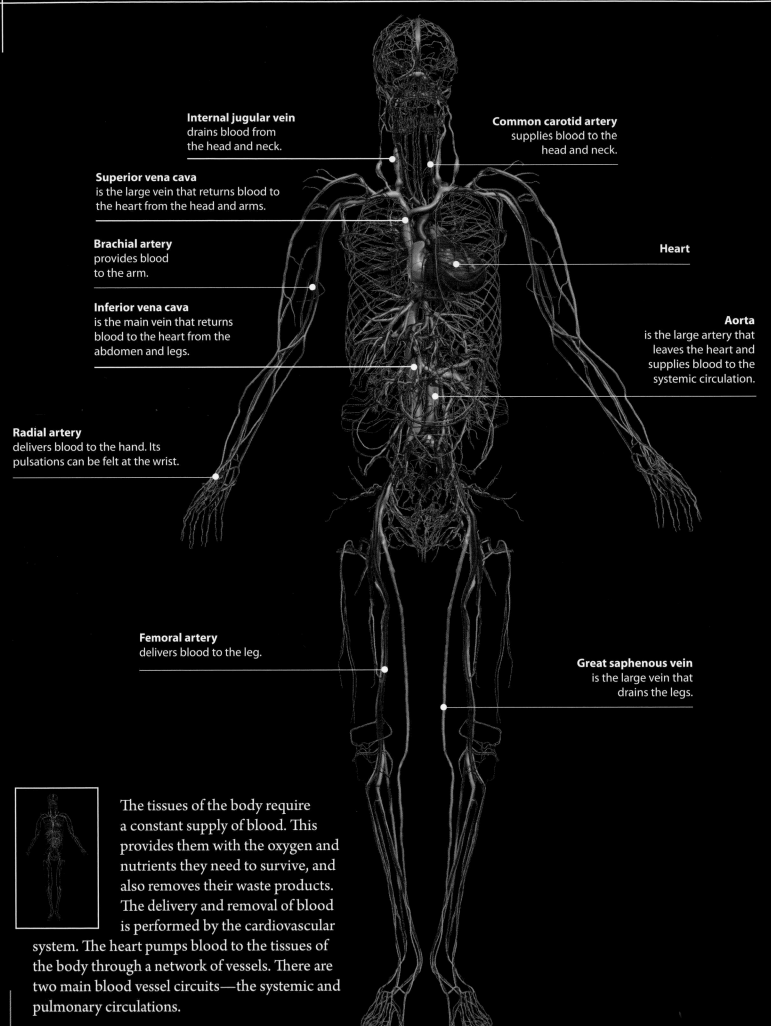

Internal jugular vein
drains blood from
the head and neck.

Common carotid artery
supplies blood to the
head and neck.

Superior vena cava
is the large vein that returns blood to
the heart from the head and arms.

Brachial artery
provides blood
to the arm.

Heart

Inferior vena cava
is the main vein that returns
blood to the heart from the
abdomen and legs.

Aorta
is the large artery that
leaves the heart and
supplies blood to the
systemic circulation.

Radial artery
delivers blood to the hand. Its
pulsations can be felt at the wrist.

Femoral artery
delivers blood to the leg.

Great saphenous vein
is the large vein that
drains the legs.

The tissues of the body require
a constant supply of blood. This
provides them with the oxygen and
nutrients they need to survive, and
also removes their waste products.
The delivery and removal of blood
is performed by the cardiovascular
system. The heart pumps blood to the tissues of
the body through a network of vessels. There are
two main blood vessel circuits—the systemic and
pulmonary circulations.

Circulatory Systems

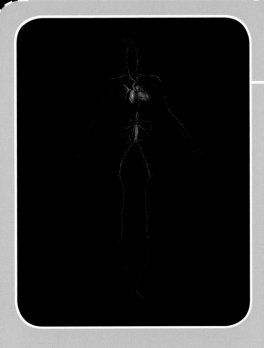

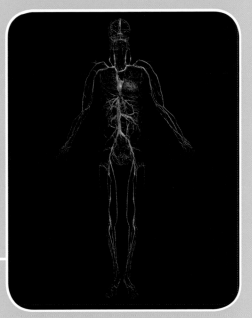

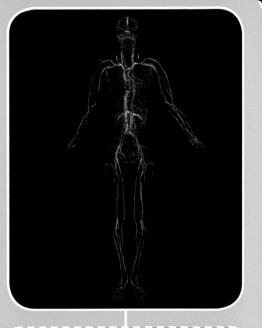

Arteries
carry blood away from the heart.

Veins
carry blood back to the heart.

Systemic Circulation
carries blood from the heart to the tissues of the body, before returning it to the heart. In the tissues, oxygen is removed from the blood and carbon dioxide added.

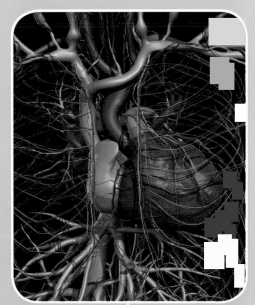

Pulmonary Circulation
carries blood from the heart to the lungs, before returning it to the heart. In the lungs, oxygen is added to the blood and carbon dioxide removed.

Portal Circulation
vessels carry nutrient rich blood from the gut for processing in the liver.

Heart
is a muscular organ which pumps blood around the network of blood vessels.

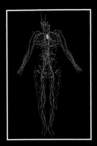

The lymphatic system has two main roles: to drain excess tissue fluid, and to fight infection. It consists of a network of lymphatic vessels, along with organs that make or contain large amounts of lymphatic tissue. These include the bone marrow, thymus, lymph nodes, and spleen. Lymphatic tissue is formed from collections of white blood cells, particularly lymphocytes and macrophages.

Thymus
is an organ present at the top of the chest. It is the place where T-lymphocytes mature. It is large in children, but shrinks as we get older.

Spleen
contains large amounts of lymphatic tissue. It monitors the blood for signs of infection.

Lymph nodes
are small bean-shaped structures found along the lymphatic vessel network. They contain large numbers of white blood cells, which monitor the lymph for signs of infection. Lymph nodes group together at certain sites, such as the neck, armpit, and groin.

Lymphatic vessels
form a network throughout the body. They prevent the buildup of fluid in the body tissues by draining and returning it to the bloodstream. The fluid present in lymphatic vessels is called lymph.

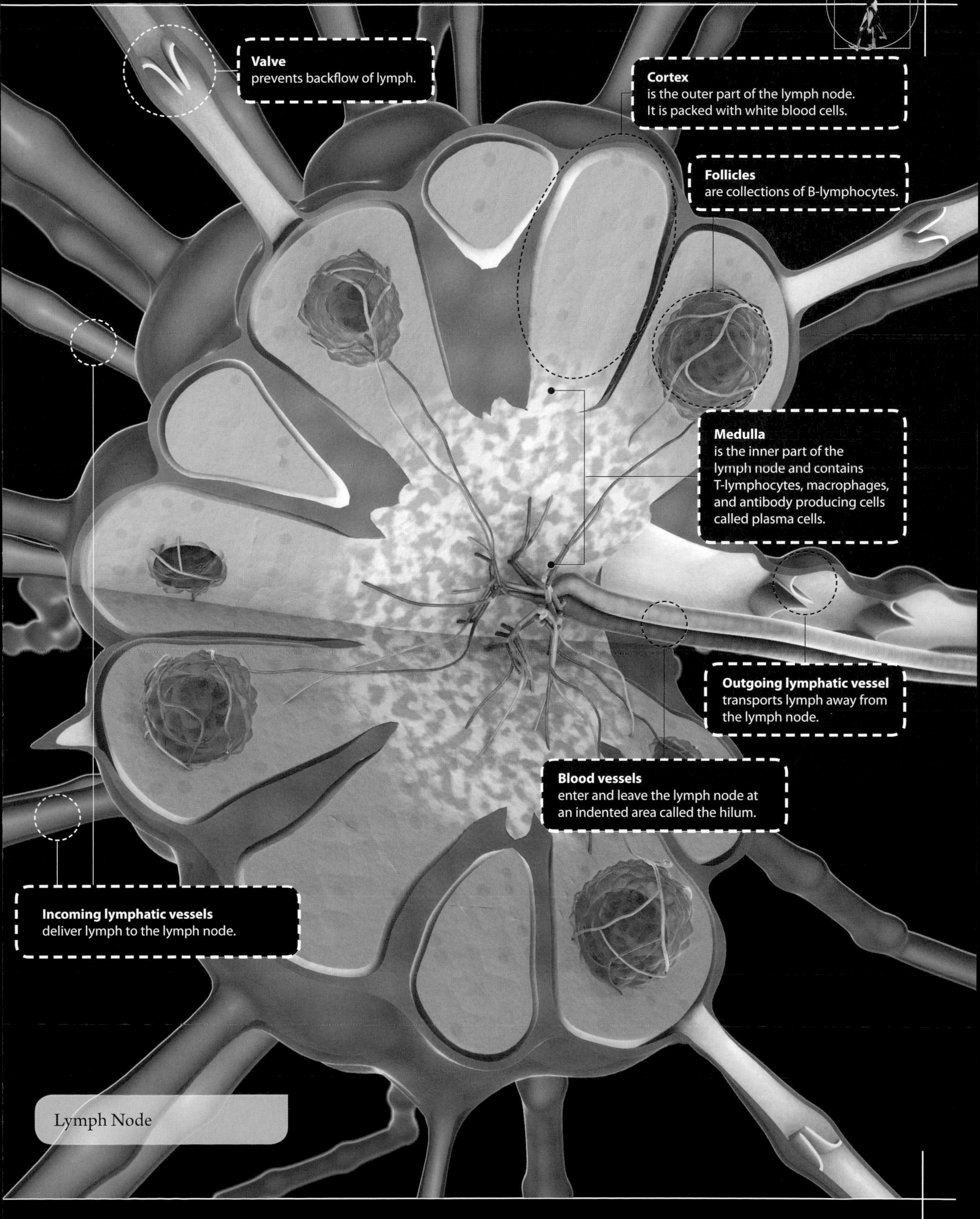

Valve
prevents backflow of lymph.

Cortex
is the outer part of the lymph node.
It is packed with white blood cells.

Follicles
are collections of B-lymphocytes.

Medulla
is the inner part of the
lymph node and contains
T-lymphocytes, macrophages,
and antibody producing cells
called plasma cells.

Outgoing lymphatic vessel
transports lymph away from
the lymph node.

Blood vessels
enter and leave the lymph node at
an indented area called the hilum.

Incoming lymphatic vessels
deliver lymph to the lymph node.

Lymph Node

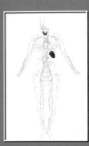

Immunity is the ability of the body to prevent and fight infections, caused by various organisms or germs. Working together, the cells of the lymphatic system play a key role in immunity. They monitor body fluids (blood and lymph) for the presence of abnormal proteins, called antigens, which indicate infection or cell damage. If detected, they respond by destroying or inactivating the infectious organism or damaged cell. This is known as an immune response.

Lymphatic System

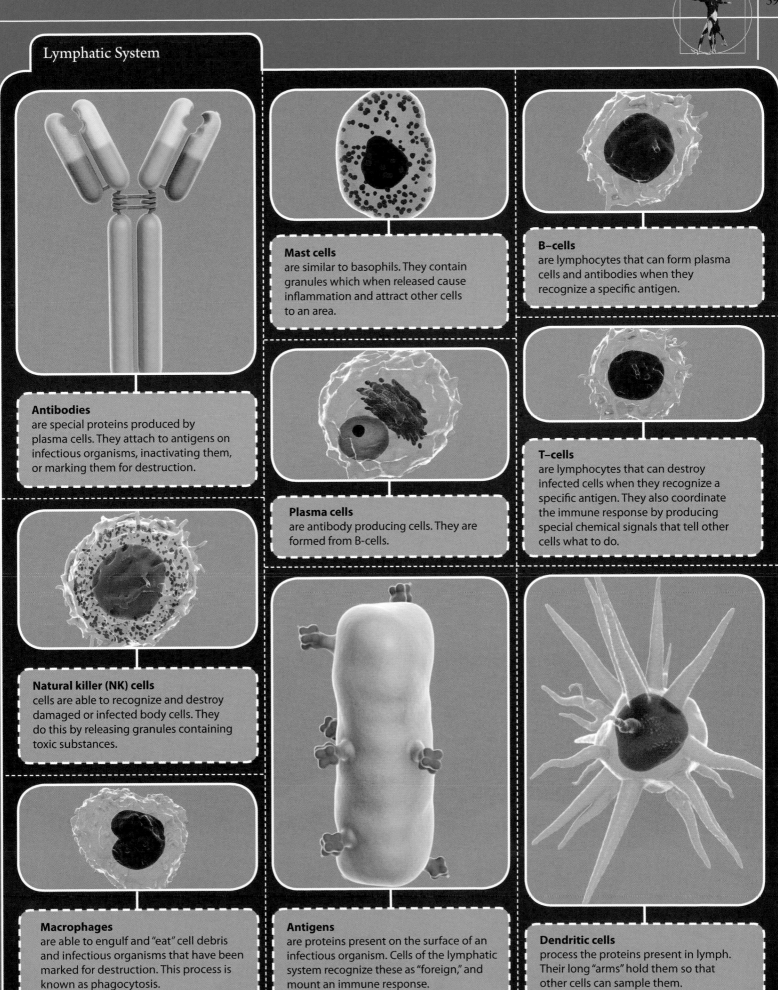

Antibodies
are special proteins produced by plasma cells. They attach to antigens on infectious organisms, inactivating them, or marking them for destruction.

Natural killer (NK) cells
cells are able to recognize and destroy damaged or infected body cells. They do this by releasing granules containing toxic substances.

Macrophages
are able to engulf and "eat" cell debris and infectious organisms that have been marked for destruction. This process is known as phagocytosis.

Mast cells
are similar to basophils. They contain granules which when released cause inflammation and attract other cells to an area.

Plasma cells
are antibody producing cells. They are formed from B-cells.

Antigens
are proteins present on the surface of an infectious organism. Cells of the lymphatic system recognize these as "foreign," and mount an immune response.

B–cells
are lymphocytes that can form plasma cells and antibodies when they recognize a specific antigen.

T–cells
are lymphocytes that can destroy infected cells when they recognize a specific antigen. They also coordinate the immune response by producing special chemical signals that tell other cells what to do.

Dendritic cells
process the proteins present in lymph. Their long "arms" hold them so that other cells can sample them.

The nervous system consists of the brain, spinal cord, and nerves. It allows us to see, hear, smell, taste, touch, move, feel, remember, and much more. To do all this requires the rapid, coordinated transmission of multiple signals, between different parts of the body. This is possible through the billions of specialized nerve cells (neurons) that conduct electrical signals, known as nerve impulses.

Chemical Synapse

The junction of a neuron with another cell (neuron, muscle or glandular cell) is called a synapse. A chemical synapse uses a neurotransmitter substance to pass the nerve impulse on.

Presynaptic neuron
delivers the nerve impulse.

Ion channels
open in response to either nerve impulses or the presence of a neurotransmitter.

Postsynaptic cell
receives the nerve impulse.

Synaptic vesicles
store the neurotransmitter before it is released.

Mitochondria

Synaptic cleft
is the gap between the two cells.

Neurotransmitters
are chemicals released from the end of a presynaptic neuron when a nerve impulse arrives. It crosses the synaptic cleft and binds to ion channels, which trigger a nerve impulse in the postsynaptic cell.

Microanatomy of a Neuron

Nerves
are cordlike structures that transmit nerve impulses to and from the spinal cord and brain. They contain bundles of axons called nerve fascicles.

Nerve fascicles
contain bundles of nerve fibers.

Neurons

Neurons are classified according to the number of fibers entering and leaving their cell body.

Unipolar

Bipolar

Multipolar

Axons
are thin fibers, which carry nerve impulses to other neurons or cells.

Cell body
contains the neuron's nucleus.

Dendrites
are relatively short, thin fibers that receive impulses from other neurons.

Neuroglia

Neuroglia are cells that provide structural and nutritional support to the neurons.

Oligodendrocytes
produce myelin sheaths for neurons within the brain and spinal cord.

Astrocytes
provide support and nutrition to neurons within the brain and spinal cord.

Microglia
help fight infection within the brain and spinal cord.

Ependymal cells
produce and monitor the fluid that circulates around the brain and spinal cord.

Nerve fibers
are single nerve axons, surrounded by a sheath of connective tissue.

Myelin sheath
insulates the axons of some neurons, allowing them to conduct impulses faster.

Axon

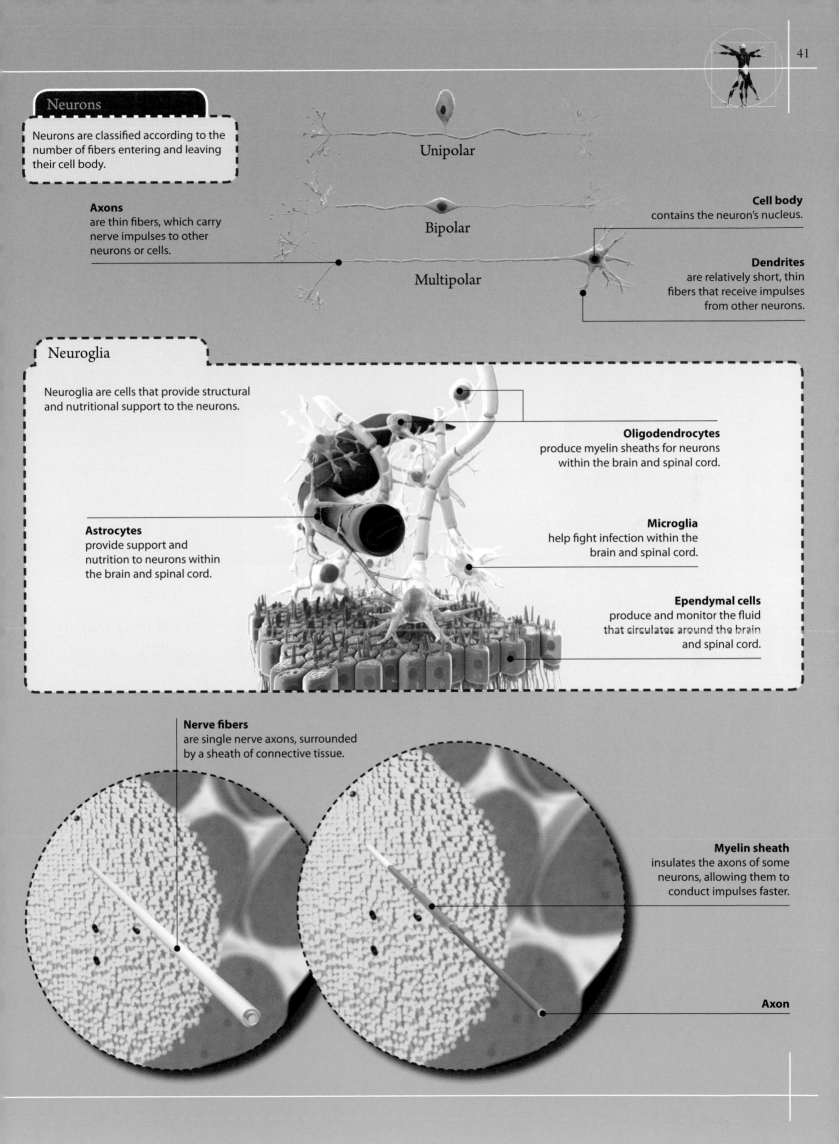

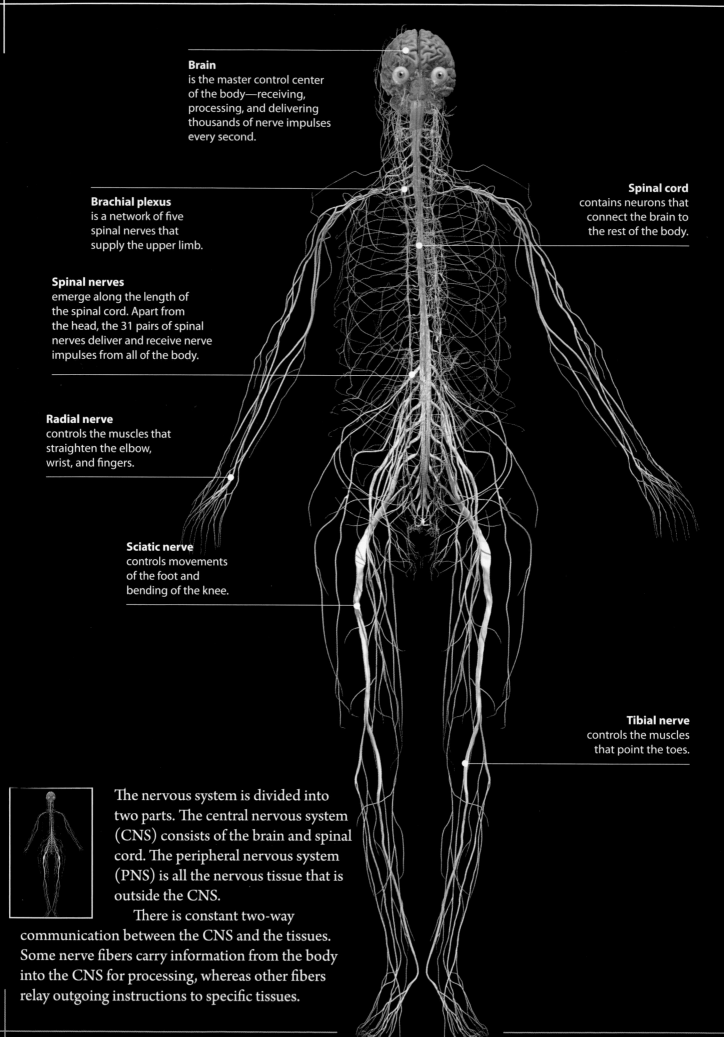

Brain
is the master control center of the body—receiving, processing, and delivering thousands of nerve impulses every second.

Spinal cord
contains neurons that connect the brain to the rest of the body.

Brachial plexus
is a network of five spinal nerves that supply the upper limb.

Spinal nerves
emerge along the length of the spinal cord. Apart from the head, the 31 pairs of spinal nerves deliver and receive nerve impulses from all of the body.

Radial nerve
controls the muscles that straighten the elbow, wrist, and fingers.

Sciatic nerve
controls movements of the foot and bending of the knee.

Tibial nerve
controls the muscles that point the toes.

The nervous system is divided into two parts. The central nervous system (CNS) consists of the brain and spinal cord. The peripheral nervous system (PNS) is all the nervous tissue that is outside the CNS.

There is constant two-way communication between the CNS and the tissues. Some nerve fibers carry information from the body into the CNS for processing, whereas other fibers relay outgoing instructions to specific tissues.

Brain

Cerebrum
is the largest part of the brain. It contains billions of neurons that are involved with tasks such as processing, thought, and memory.

Gyri
are the folds of the cerebrum.

Sulci
are the grooves between the gyri.

Cerebellum
is involved with balance and coordination.

Brainstem
connects the brain with the spinal cord. It contains neurons which control vital functions, such as breathing.

Spinal Cord

Grey matter
contains the cell bodies of neurons.

White matter
contain the axons of neurons running up and down the spinal cord.

Posterior roots
are where incoming sensory information enters the spinal cord.

Spinal nerves
carry both motor and sensory information.

Anterior roots
are where outgoing instructions leave the spinal cord.

ENDOCRINE SYSTEM

The endocrine system consists of a number of glandular organs spread throughout the body. They control what happens inside the body by releasing chemical messengers, called hormones, into the bloodstream. These hormones act on specific tissues in the body, causing various changes to occur.

Thyroid gland
is located at the front of the neck. It produces thyroid hormones, which increase the rate of metabolism within cells.

Parathyroid glands
are four tiny glands found at the back of the thyroid gland. They produce parathyroid hormone and calcitonin, which control the body's calcium levels.

Thymus
produces hormones that help lymphocytes to mature.

Suprarenal glands
are found above the kidneys. They produce various steroid hormones in their outer part (cortex) that regulate salt levels and alter cell metabolism. The middle of the gland (medulla) produces adrenaline that helps the body respond to stressful situations.

Ovaries
are paired glands which produce the female sex hormones, estrogen and progesterone. These hormones regulate the monthly menstrual periods, as well as causing female features to develop.

Hypothalamus
is part of the brain that works with the pituitary gland to control various endocrine organs in the body. It links the nervous and endocrine systems.

Pituitary gland
is a pea-sized gland found near the base of the brain. It receives signals from the hypothalamus, which regulate the release of eight hormones. These either act directly on body tissues or on other endocrine organs.

Pancreas
secretes various hormones, the two most important being insulin and glucagon. They control the levels of sugar (glucose) within the blood.

Testes
are paired glands that produce the male sex hormone testosterone. This assists the production of sperm, as well as causing male features to develop.

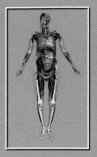

The respiratory system consists of a series of dividing passageways that carry air in and out of the numerous, tiny, inflatable sacs in the lungs. Their function is to provide a constant supply of "fresh" oxygen rich air from the environment, while removing carbon dioxide from the body.

Body cells and tissues need oxygen, to produce energy from nutrients. A waste product of this process is carbon dioxide, which if it accumulates would poison the cells. Blood carries oxygen to the tissues and removes carbon dioxide. The lungs provide a site where blood from the tissues can get rid of its carbon dioxide, and be recharged with oxygen.

Mouth and nose provide openings that connect the respiratory system to the fresh air outside the body. The mouth and nose are connected at the back of the throat.

Left main bronchus carries air to and from the left lung .

Trachea or windpipe, splits at its lower end into the right and left main bronchi.

Diaphragm is the main muscle that moves air in and out of the lungs.

Nose
warms, humidifies, and
filters air drawn into it.

Larynx (voice box)
connects the throat and the
trachea. It contains folds of tissues
called the vocal cords, which can
use passing air to vibrate and
produce sounds for speech.

Lungs
contain a series of branching
passageways that eventually lead
to small air sacs. These are where
oxygen can be added to the blood
and carbon dioxide is removed,
in a process called gas exchange.

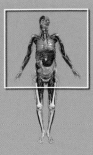

The cells of the body require nutrients for energy. This is usually provided by food from the diet. However, food needs breaking down into small molecules to allow it to be used by the body. This process of breaking down food and absorbing it is the function of the digestive system.

The main part of the digestive system is a long tube that runs from the mouth to the anus, known as the alimentary canal. In addition, some organs attach to the alimentary canal, producing substances that help the digestive processes.

Esophagus
is a muscular tube that transports food from the mouth to the stomach.

Stomach
is a sac-like organ. It contains strong acid and enzymes which break down food.

Rectum
signals to the body when feces are building up and need to be removed.

Anal canal
is the final part of the alimentary canal. It has strong circular muscles that prevent feces from being released until it is convenient.

Mouth
contains the teeth, which cut
and chew food, breaking it into
smaller pieces for processing.

Liver
is a large organ that produces bile
for storage by the gallbladder. It also
processes and stores many nutrients once
they are absorbed into the bloodstream.

Gallbladder
is a hollow organ that stores
bile produced by the liver. It
releases it into the small bowel
to help break down fats.

Small bowel
receives stomach contents, and
continues to digest them using
various enzymes. It also absorbs
small nutrient molecules from
the alimentary canal.

Large bowel
forms waste (feces) by
removing water from
any undigested material.

Kidneys
are paired organs that lie
on either side of the spine.
They filter the blood and
produce urine.

Male Urinary System

Female Urinary System

Ureters
are muscular tubes, that
carry urine from the
kidneys to the bladder.

Bladder
is a hollow organ with muscular
walls. It can expand to store
various amounts of urine.

Urethra
is the tube that carries urine
from the bladder to the
outside. Males have longer
urethras than females.

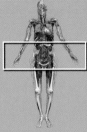

The kidneys, ureters, bladder, and urethra
make up the urinary system. The role
of the urinary system is to remove waste
substances from the body. It does this by
continually filtering and cleaning the blood,
producing a waste fluid called urine. This can
be stored, until an appropriate time and place
is found to get rid of it from the body, in a process called
urination. The urinary system also helps control the
levels of salts and fluid in the body.

Kidney

Cortex
is the outer part of the kidney where blood is filtered.

Calyces
are the "horn-shaped" tubes that carry urine to the renal pelvis.

Medulla
is the inner part that determines how concentrated the urine is. It drains urine into the calyces.

Renal pelvis
is the wide area, which funnels urine from the calyces into the ureter.

Ureter

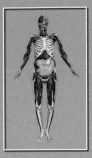

The integumentary system consists of the skin, along with accessory structures such as hairs, glands, and nails. It provides the body with a waterproof, germ-resistant barrier. It also helps control body temperature, and allows us to touch and feel objects around us.

Did you know?

The total weight of the skin of an average adult human is 9-11lbs.

The average growth rate of fingernails is 1 millimeter per week. The toenails grow slower than this.

Skin

Hair
is made up of dead epidermal cells held together by special proteins. They are present on most body surfaces but are concentrated on the top of the head.

Epidermis
is the outermost layer of skin. It forms a protective, waterproof barrier.

Dermis
is the innermost layer of skin. It is made up of tough, flexible fibrous tissue, and contains blood vessels and accessory structures.

Sebaceous glands
produce an oily substance called sebum, which helps moisturize the skin and hair.

Sweat glands
help cool the body.

Nerve fibers
detect different sensations like pressure, touch, pain and vibration.

Hypodermis
is a layer of fatty tissue underneath the dermis that helps attach the skin to the body.

Blood vessels
help to control body temperature as well as supplying the skin tissues.

Hair

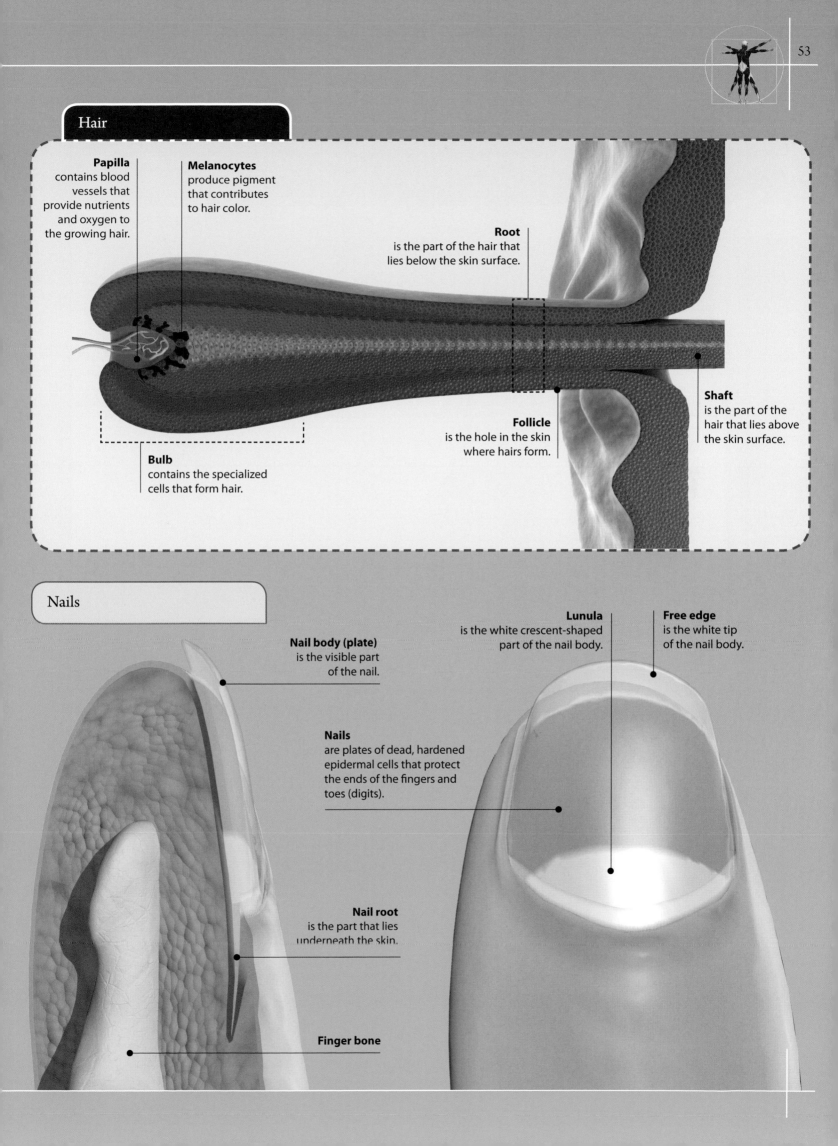

Papilla
contains blood vessels that provide nutrients and oxygen to the growing hair.

Melanocytes
produce pigment that contributes to hair color.

Root
is the part of the hair that lies below the skin surface.

Shaft
is the part of the hair that lies above the skin surface.

Follicle
is the hole in the skin where hairs form.

Bulb
contains the specialized cells that form hair.

Nails

Nail body (plate)
is the visible part of the nail.

Lunula
is the white crescent-shaped part of the nail body.

Free edge
is the white tip of the nail body.

Nails
are plates of dead, hardened epidermal cells that protect the ends of the fingers and toes (digits).

Nail root
is the part that lies underneath the skin.

Finger bone

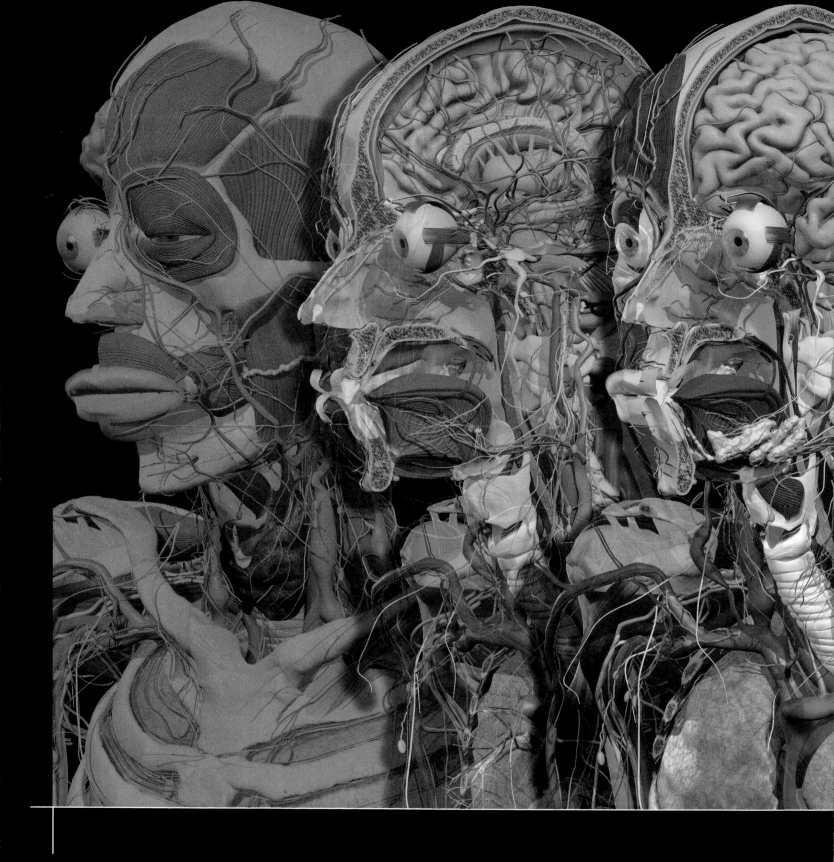

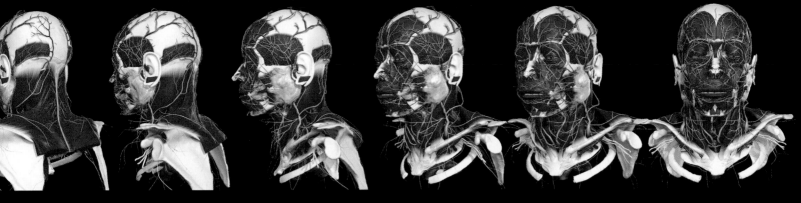

The head contains some of our most important and complex organs, including the brain, eyes, ears, nose, and tongue.

The bony skull provides protection to these delicate structures, as well as giving shape to our face. The mouth and nose provide openings to the outside world, allowing us to take in food, water, and air, while the numerous muscles help us to speak, chew, and show whether we are happy or sad.

The neck supports and moves the head. It is also the communication channel between the structures of the head and the rest of the body. Running through the neck are the spinal cord, the windpipe, the esophagus, and many large blood vessels.

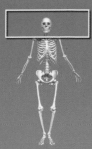

The cranium, or skull, is made up of a number of individual bones that are fixed together at fused joints called sutures. It provides protection for delicate structures, such as the brain and the eyes, and also provides a solid base for muscles of the head and neck to attach to. The area inside the skull is called the cranial cavity, and houses the brain. Within the bones there are holes, called foramen, which allow the passage of nerves and blood vessels to and from the cranial cavity.

Suture line

Orbits
or eye sockets, provide almost complete protection for the eyes.

Optic canal
is an opening at the back of the orbit. The optic nerve passes through this, connecting the brain to the eye.

Nasal cavity
is one of the openings of the respiratory system. It contains special nerves that allow the detection of odors and smells.

Mandible
or jaw bone

Axial section

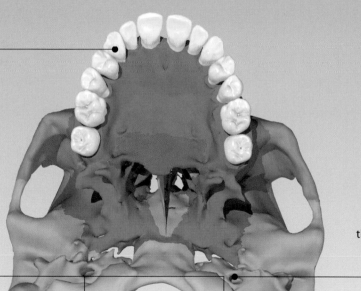

Posterior cranial fossa

Middle cranial fossa

Anterior cranial fossa
is the highest and furthest forward of three distinct recessed areas within the cranial cavity.

Cribriform plate
contains small holes that connect the nasal cavity to the brain. Specialized nerves that carry the sense of smell travel through these holes.

Teeth
are firmly attached to the jaw bones. They have a tough outer coating of enamel that allows them to cut and chew different types of food.

Occipital condyles
are where the skull joins the vertebral (spinal) column.

Carotid canal
is a passage that carries the internal carotid artery into the cranial cavity to supply the brain.

Foramen magnum
is the large hole at the base of the skull. The spinal cord passes through this, connecting the brain to the rest of the body.

Base

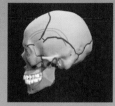
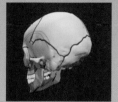
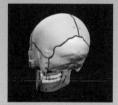
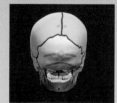
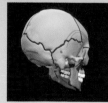
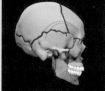
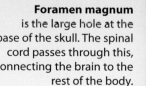

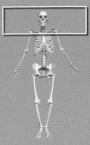

The cranium, or skull bone, is made up of twenty-two separate bones. Of these, eight cranial bones form the cranial cavity which houses the brain. The remaining fourteen bones form the face, and are known as facial bones. Most of the bones are fixed in place and are attached to each other by irregular joint lines called sutures.

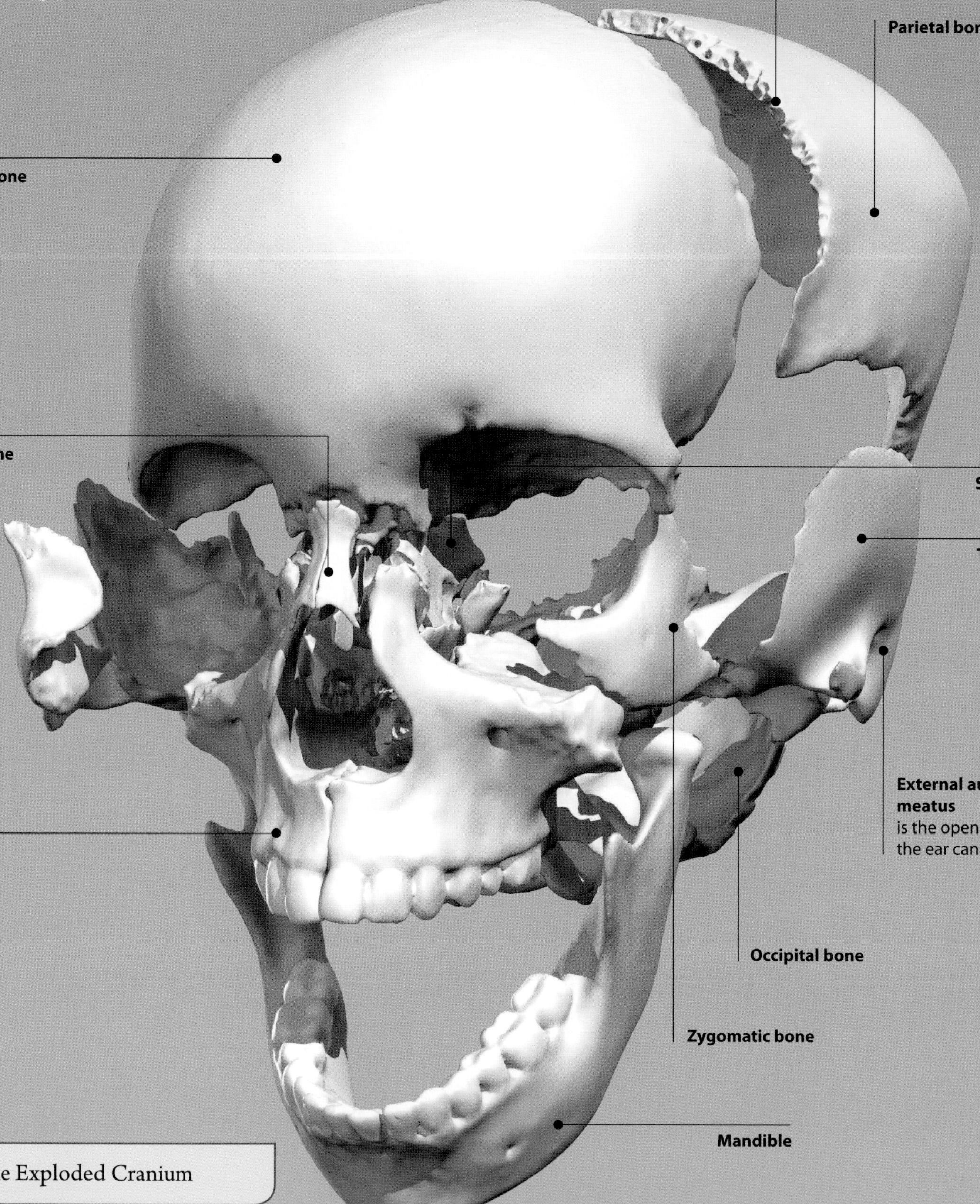

Sutures
are solid immovable joints that form between the skull bones. The irregular jagged edges fix the bones in place.

Parietal bones

Frontal bone

Nasal bone

Sphenoid bone

Temporal bone

Maxilla

External auditory meatus
is the opening of the ear canal.

Occipital bone

Zygomatic bone

Mandible

The Exploded Cranium

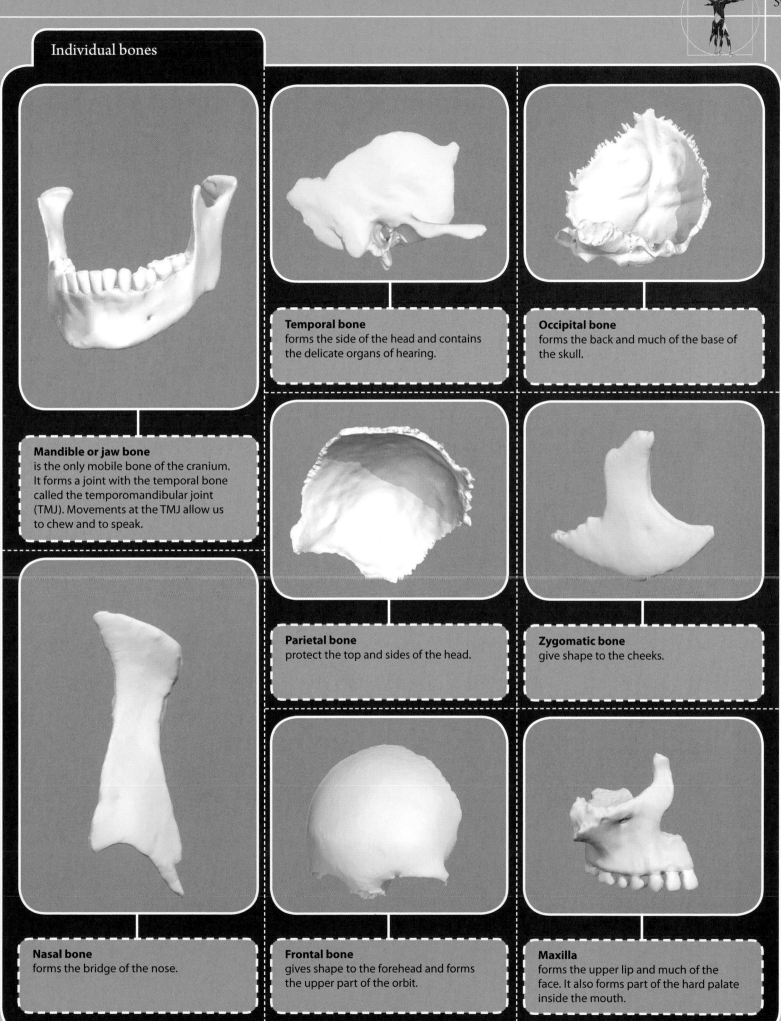

Individual bones

Temporal bone
forms the side of the head and contains the delicate organs of hearing.

Occipital bone
forms the back and much of the base of the skull.

Mandible or jaw bone
is the only mobile bone of the cranium. It forms a joint with the temporal bone called the temporomandibular joint (TMJ). Movements at the TMJ allow us to chew and to speak.

Parietal bone
protect the top and sides of the head.

Zygomatic bone
give shape to the cheeks.

Nasal bone
forms the bridge of the nose.

Frontal bone
gives shape to the forehead and forms the upper part of the orbit.

Maxilla
forms the upper lip and much of the face. It also forms part of the hard palate inside the mouth.

The muscles of facial expression allow us to show how we are feeling through the movements and appearance of the face. They are an unusual group of muscles, as instead of just connecting to bones, one of their ends is attached to the skin. This means that when they contract, they move the overlying skin, generating different facial expressions. We use these muscles to show a wide range of emotions—from happiness and surprise, to anger and sadness. Their names often give clues to their action, or what they are attached to. They are all supplied by the facial nerve.

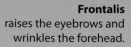

Frontalis
raises the eyebrows and wrinkles the forehead.

Orbicularis oculi
closes the eyes.

Procerus
wrinkles up the nose.

Zygomaticus major
pulls the corners of the mouth upwards and to the side, forming a smile.

Levator anguli oris
raises the corners of the mouth.

Buccinator
pulls the cheeks against the teeth, as in sucking or whistling.

Risorius
pulls the edges of the mouth to the sides, as in grimacing.

Depressor labii inferioris
pulls the lower lip down.

Platysma
wrinkles the skin of the neck and pulls the lower lip down.

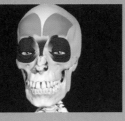

Nasalis
widens the openings
of the nose, "flaring"
the nostrils.

Levator labii superioris
raises the upper lip.

Orbicularis oris
closes the mouth, and
is also involved with
pursing the lips for a kiss.

Depressor anguli oris
pulls down the corners
of the mouth.

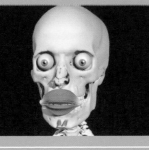
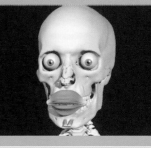
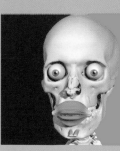

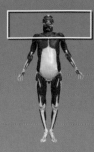

The upper and lower jaw contains teeth, which are used to cut and chew food. Four strong muscles work together to produce movements necessary for chewing and are known as the muscles of mastication (chewing). They move the lower jaw (mandible) up and down, backward and forward, and from side to side.

Movements of the mandible take place at the temporomandibular joint (TMJ). This joint is formed where the lower jaw bone meets the skull, just in front of the ear.

Temporalis
is a broad muscle on the side of the head. It moves the mandible up and backward.

Temporomandibular joint

Masseter
is the most powerful of the muscles of mastication. It moves the mandible up, bringing the teeth together.

Mandible

Buccinator
is a muscle of facial expression that stops food accumulating between the cheeks and the teeth.

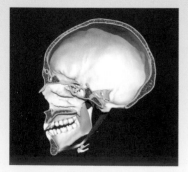
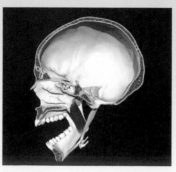
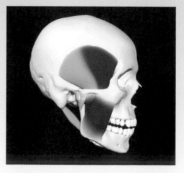
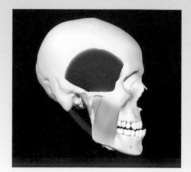

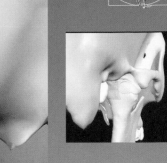

Temporomandibular Joint

Articular disks
cushion the joint and allow the
different movements to take place.

Temporal bone
is one of the bones of the skull.

External auditory
meatus is the opening
of the ear canal.

Joint capsule
covers the joint.

Stylomandibular ligament
is a cord of fibrous tissue that
attaches the base of the skull to
the mandible. It stops the jaw
from opening too wide.

Mandible
or lower jaw bone moves during
chewing (mastication).

Lateral pterygoid
moves the jaw forwards
and from side to side.

Medial pterygoid
closes the mouth, and along with
the lateral pterygoid moves the
mandible from side to side.

Teeth

Mandible

Buccinator

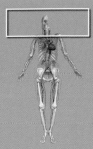

Neurovascular structures are the arteries, veins, and nerves that supply, drain, and innervate tissues in a particular body region. In the head, they form large branching networks that cover the face, scalp, nasal, and oral cavities, along with the brain and eyes.

Arteries of the Head

Superficial temporal artery
runs just in front of the ear. It supplies the sides and top of the head.

Maxillary artery
supplies the mouth and nose.

Facial artery
crosses the lower jawbone and runs diagonally across the face. It supplies the cheeks, lips, and nose.

External carotid artery
is the branch of the common carotid artery that supplies the tissues of the face and scalp.

Common carotid arteries
provide most of the blood to the tissues of the head. It divides into two large branches.

Internal carotid artery
is the branch of the common carotid artery that supplies the brain and eyes.

Nerves of the Head

Facial nerve
branches carry motor instructions to the muscles of facial expression.

Supraorbital and supratrochlear nerves
carry sensation from above the eye and the middle of the forehead.

Lesser occipital nerve
carries sensation from the back of the neck.

Alveolar nerves
carry sensation from the teeth.

Veins of the Head

Facial vein
drains blood from the cheeks, lips, and nose.

Posterior auricular vein
drains blood from behind the ear.

Internal jugular vein
drains blood from the brain, face, and scalp.

External jugular vein
drains blood from the face and scalp.

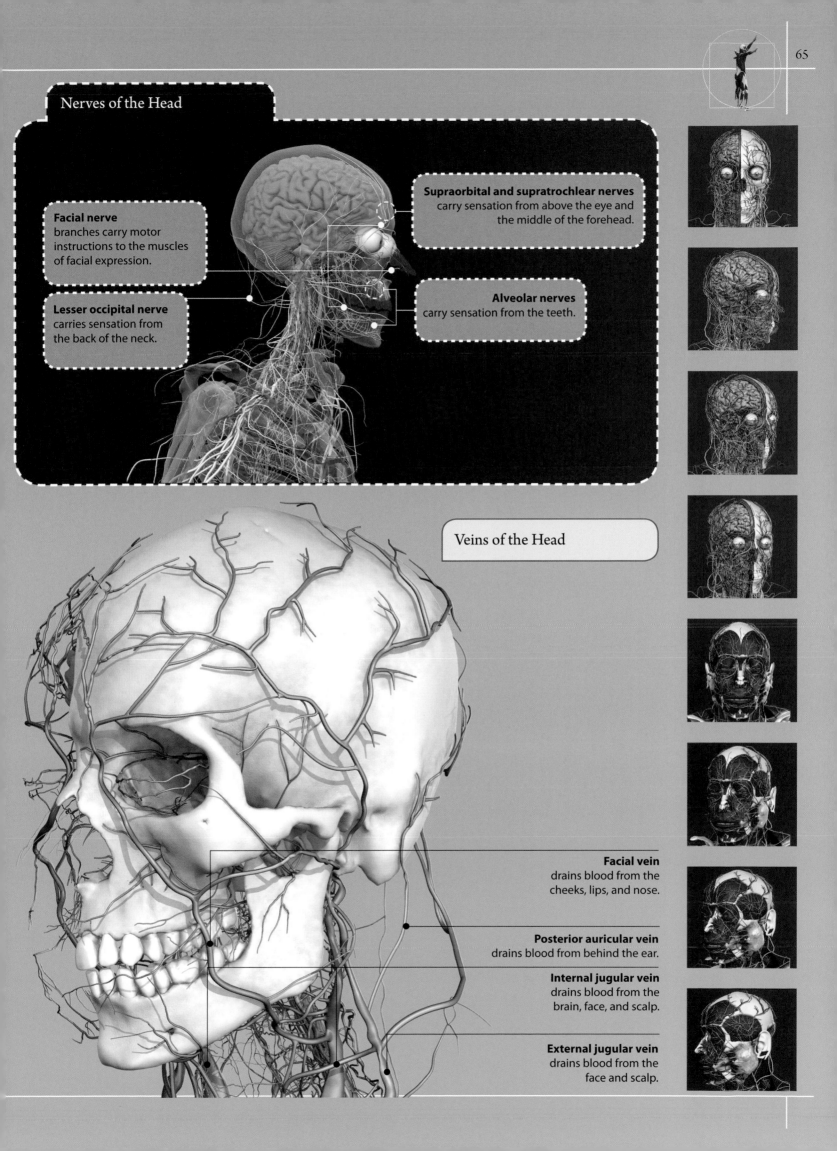

CRANIAL NERVES I-VI

The cranial nerves are so named as they all emerge directly out of the brain, within the cranial cavity. There are twelve pairs of cranial nerves (CN) that are numbered using Roman numerals. Between them, they provide motor and sensory innervation to the head and neck regions. They all pass through various holes in the skull to reach the areas they supply. The first six cranial nerves allow us to smell, to see, to move our eyes, as well as providing sensation to the face, and motor control to the muscles of mastication.

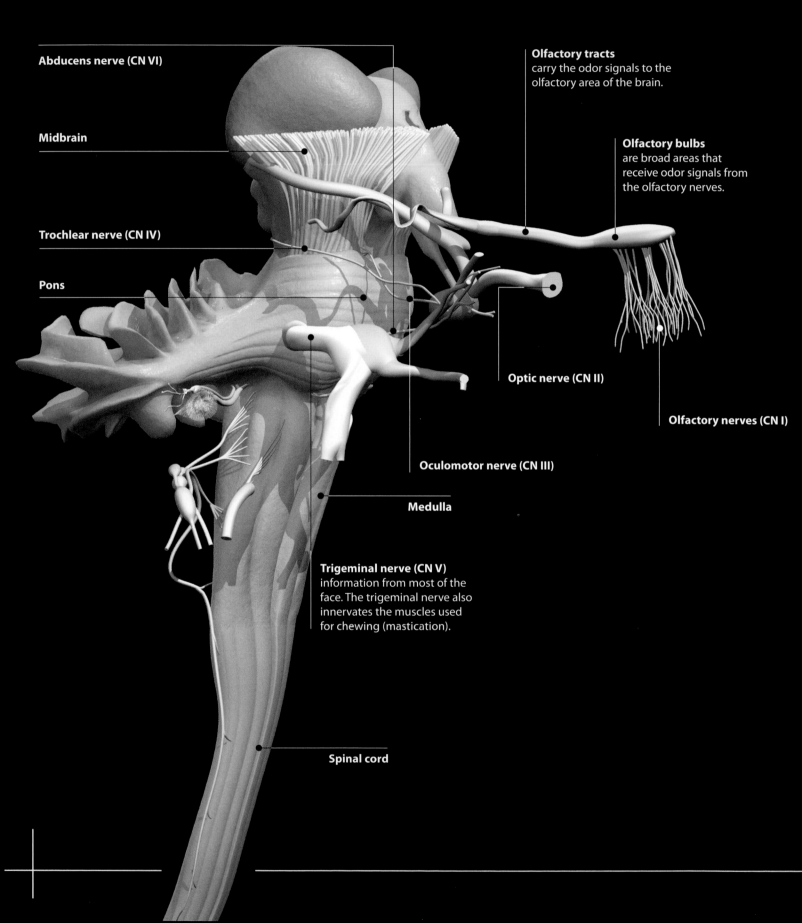

Abducens nerve (CN VI)

Midbrain

Trochlear nerve (CN IV)

Pons

Olfactory tracts
carry the odor signals to the olfactory area of the brain.

Olfactory bulbs
are broad areas that receive odor signals from the olfactory nerves.

Optic nerve (CN II)

Olfactory nerves (CN I)

Oculomotor nerve (CN III)

Medulla

Trigeminal nerve (CN V)
information from most of the face. The trigeminal nerve also innervates the muscles used for chewing (mastication).

Spinal cord

Cranial Nerves I-VI

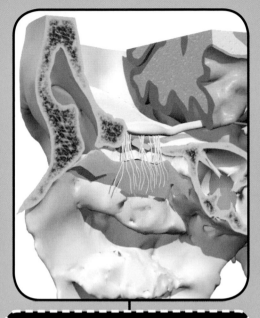

Olfactory nerves (CN I)
detect odors and allow us to smell. They are located at the top of the inside of the nose.

Optic nerve (CN II)
carries visual information from the back of the eyes to the brain to allow us to see.

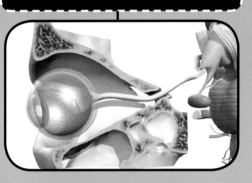

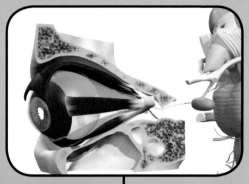

Oculomotor nerve (CN III)
controls most of the muscles that move the eyeball. It also opens the eyelids, and makes the pupils of the eyes smaller.

Trigeminal nerve (CN V)
has three main branches; the ophthalmic, maxillary, and mandibular divisions. Between them they carry sensory information from most of the face. The trigeminal nerve also innervates the muscles used for chewing (mastication).

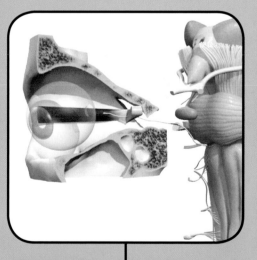

Abducens nerve (CN VI)
controls the muscle that makes the eye look to the side.

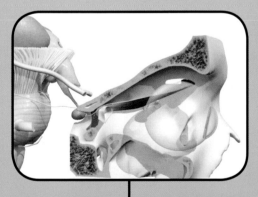

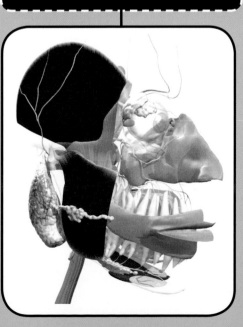

Trochlear nerve (CN IV)
controls a muscle that helps rotate the eye down and out.

Cranial nerves (CN) VII to XII have a wide variety of functions. They control the muscles of facial expression, and allow us to hear, keep our balance, shrug our shoulders, taste, swallow, and cry. By traveling outside the head and neck regions, the vagus nerve helps control the heart and gastrointestinal system. These nerves can even make our mouths water, by supplying the salivary glands.

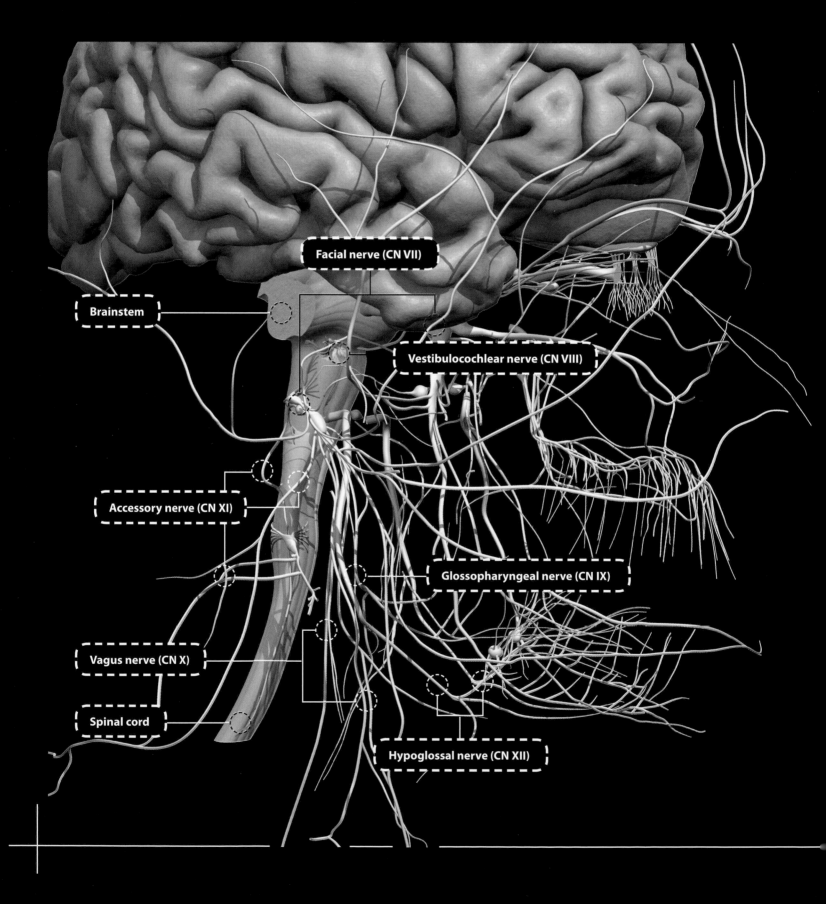

Facial nerve (CN VII)

Brainstem

Vestibulocochlear nerve (CN VIII)

Accessory nerve (CN XI)

Glossopharyngeal nerve (CN IX)

Vagus nerve (CN X)

Spinal cord

Hypoglossal nerve (CN XII)

Cranial Nerves VII-XII

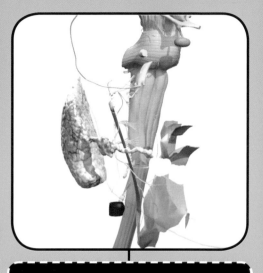

Vestibulocochlear nerve (CN VIII)
has two branches, allowing us to hear
(cochlear part) and to keep our balance
(vestibular part).

Facial nerve (CN VII)
branches innervate the muscles of the
face, allowing us to smile, frown, and
close our eyes. It carries taste sensation
from the front of the tongue as well as
supplying the lacrimal (tear) and salivary
glands.

Glossopharyngeal nerve (CN IX)
is predominantly a sensory nerve,
carrying information from the back of
the throat (pharynx) and deep structures
of the neck. These signals allow us to
swallow properly, as well as controlling
the blood pressure.

Accessory nerve (CN XI)
has branches from both the brainstem
and the spinal cord. It innervates muscles
in the neck and upper back that shrug
the shoulders and turn the head.

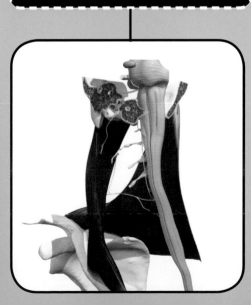

Vagus nerve (CN X)
innervates muscles of the throat and
larynx (voicebox) that allow us to
swallow, speak, and cough. It is known as
the "wandering" nerve, as it travels down
into the chest and abdomen, where it
helps control the cardiovascular and
gastrointestinal systems.

Hypoglossal nerve (CN XII)
controls the muscles that move the
tongue.

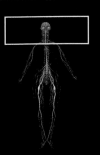

The brain is the master control center of the body, coordinating the activities of the body systems. It allows the body to move, feel, think, remember, and speak, performing thousands of complex calculations every second. It is a relatively small organ considering its importance, weighing just over a kilogram. The brain can be divided into distinct areas, all serving different functions.

Sulci
are the deep grooves between the gyri. Some of them are used to divide the cerebrum into lobes.

Frontal lobe
The lobes of the brain are named after the bones of the skull that cover them.

Parietal lobe

Occipital lobe

Cerebellum
lies behind the brainstem and underneath the cerebrum. It helps coordinate movements.

Temporal lobe

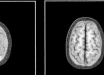

Medial Section

Corpus callosum
is a thick band of nerve fibers
that connect the right cerebral
hemisphere to the left.

Diencephalon
is the region of brain between the
brainstem and the cerebrum.

Pituitary gland
is a small pea-sized structure
connected to the base of the brain.
It controls the endocrine system.

Ventricles
are fluid-filled cavities within the brain.

Brainstem
connects the brain to the rest of
the body through the spinal cord.

The brain is separated from the bones of the skull by layers of fibrous tissue called the meninges. There are three meningeal layers: the dura mater, the arachnoid mater, and the pia mater. The dura mater is further divided into a periosteal and meningeal layer. Together, the meninges form a tough protective covering for the brain. In places the meningeal layer of the dura mater forms thickened folds, which separate parts of the brain, as well as providing support.

Cerebrospinal fluid (CSF) is produced in the ventricles. It bathes the brain, acting like a cushion, while also providing nutrition. It circulates in the subarachnoid space, found between the arachnoid and pia meningeal layers.

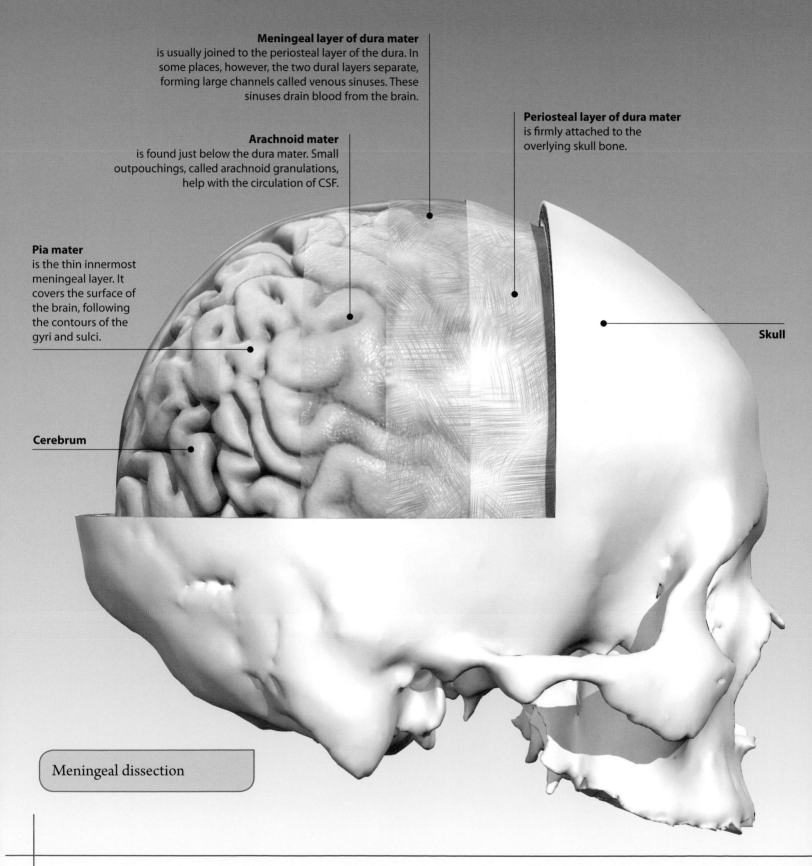

Meningeal layer of dura mater
is usually joined to the periosteal layer of the dura. In some places, however, the two dural layers separate, forming large channels called venous sinuses. These sinuses drain blood from the brain.

Periosteal layer of dura mater
is firmly attached to the overlying skull bone.

Arachnoid mater
is found just below the dura mater. Small outpouchings, called arachnoid granulations, help with the circulation of CSF.

Pia mater
is the thin innermost meningeal layer. It covers the surface of the brain, following the contours of the gyri and sulci.

Skull

Cerebrum

Meningeal dissection

Did you know?

Hydrocephalus is a serious condition that can occur if there is disruption to the circulation of CSF. The CSF builds up and puts pressure on the brain. Treatment may involve surgically diverting, or shunting, the excess CSF into the abdominal cavity.

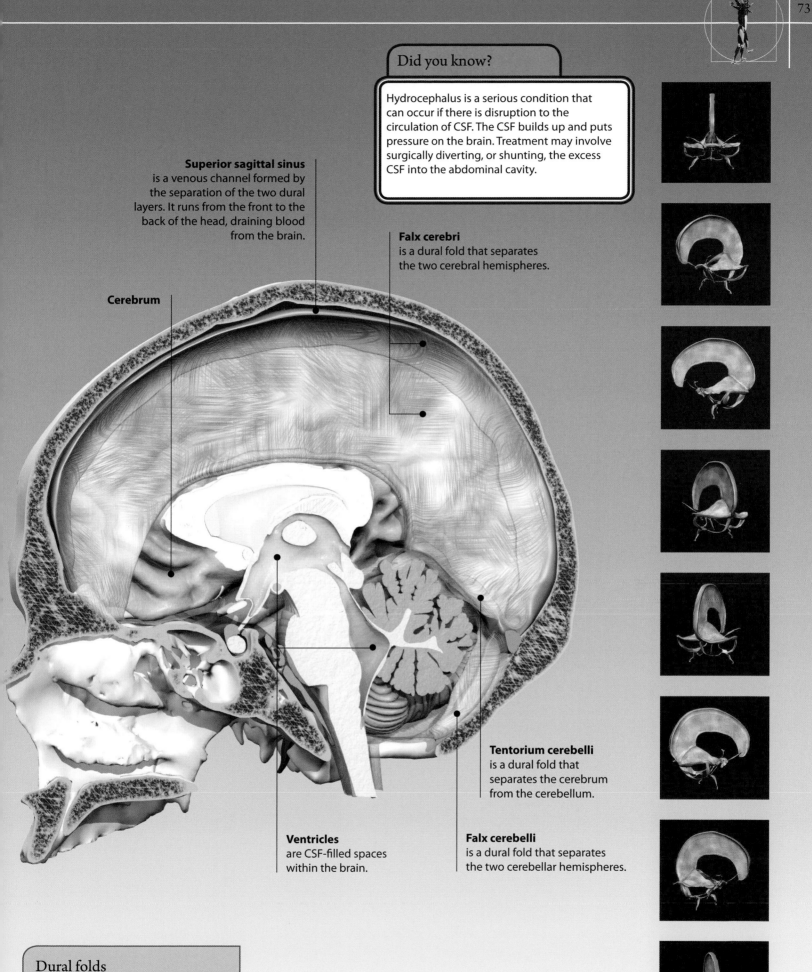

Superior sagittal sinus
is a venous channel formed by the separation of the two dural layers. It runs from the front to the back of the head, draining blood from the brain.

Falx cerebri
is a dural fold that separates the two cerebral hemispheres.

Cerebrum

Tentorium cerebelli
is a dural fold that separates the cerebrum from the cerebellum.

Ventricles
are CSF-filled spaces within the brain.

Falx cerebelli
is a dural fold that separates the two cerebellar hemispheres.

Dural folds

The brainstem connects the brain to the rest of the body. It consists of the midbrain, pons, and medulla. As well as containing all nerve fibers traveling to and from the spinal cord, the brainstem also has control centers for functions such as respiration (breathing), vomiting, and the control of blood pressure. It also contains the nuclei of the cranial nerves.

    The cerebellum, or "little brain," is located behind the brainstem. It helps to coordinate and fine tune movements, as well as contributing to balance. Via the brainstem, it has numerous connections with other areas of the brain.

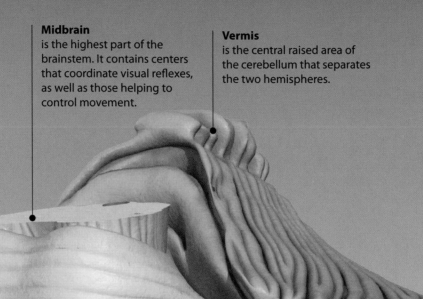

Midbrain
is the highest part of the brainstem. It contains centers that coordinate visual reflexes, as well as those helping to control movement.

Vermis
is the central raised area of the cerebellum that separates the two hemispheres.

Pons
is the expanded middle part of the brainstem. It acts like a bridge, linking different areas of the brain. Centers controlling breathing are also present here, along with cranial nerve nuclei.

Cerebellar hemispheres
are the two sides of the cerebellum.

Cerebellar hemispheres
are the two sides of the cerebellum.

Medulla
is the lowest part of the brainstem. It is continuous with the spinal cord as it leaves the skull through the foramen magnum. It contains centers that coordinate breathing and blood pressure.

Olives
are round bulges found on the front and side of the medulla. They relay signals to the cerebellum.

Cerebellum
has a tightly folded surface.

Pyramids
are two protruding columns found at the front of the medulla. They contain bundles of motor nerve fibers traveling to the rest of the body.

Brainstem & Cerebellum in detail

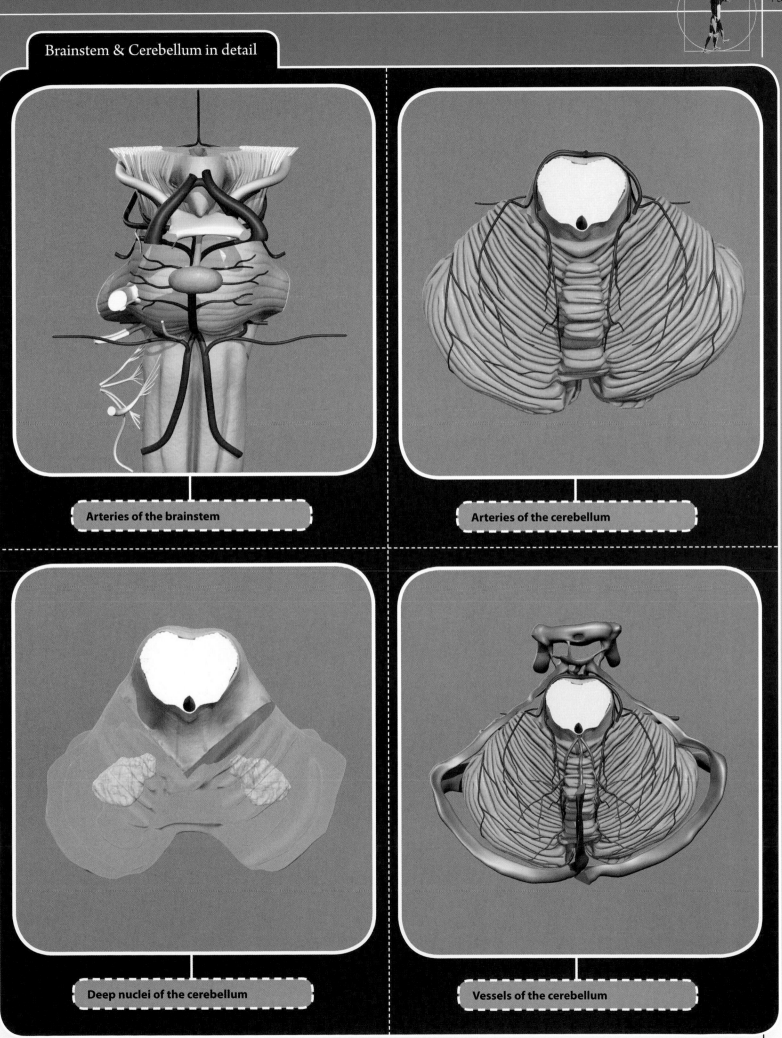

Arteries of the brainstem

Arteries of the cerebellum

Deep nuclei of the cerebellum

Vessels of the cerebellum

The ventricles are interconnected, fluid-filled cavities within the brain. Specialized blood vessels in their walls produce cerebrospinal fluid (CSF), which circulates through the ventricles, before entering the subarachnoid space to bathe the surface of the brain. The CSF provides a cushion and support for the brain, as well as nutrition. There are four ventricles in total: two lateral, one third, and one fourth ventricle.

Curved Axial Section through the Brain

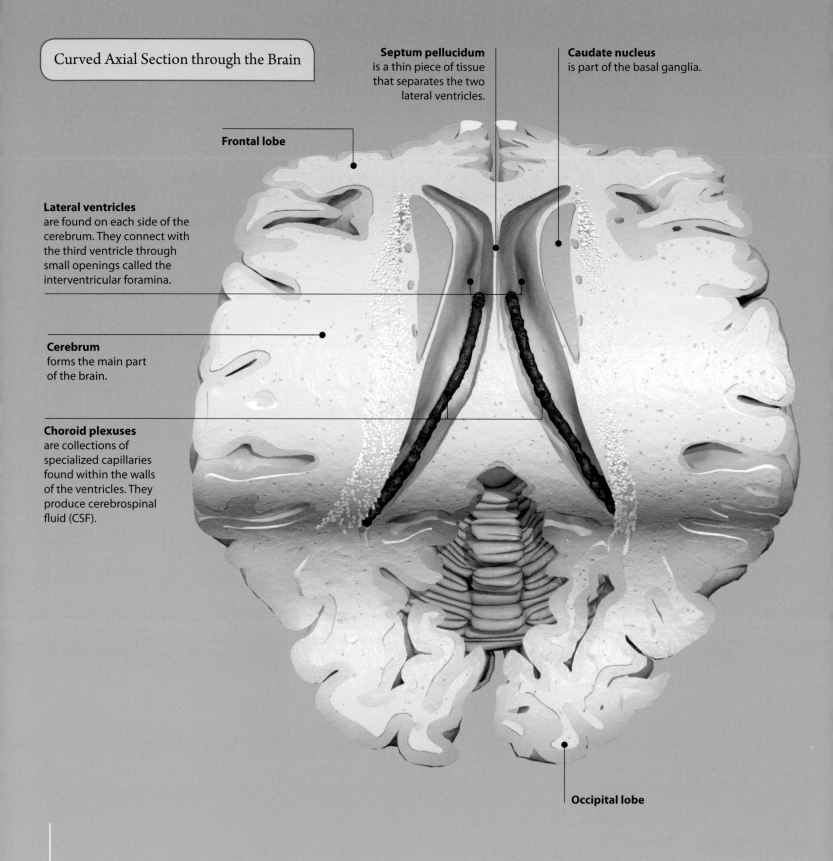

Septum pellucidum is a thin piece of tissue that separates the two lateral ventricles.

Caudate nucleus is part of the basal ganglia.

Frontal lobe

Lateral ventricles are found on each side of the cerebrum. They connect with the third ventricle through small openings called the interventricular foramina.

Cerebrum forms the main part of the brain.

Choroid plexuses are collections of specialized capillaries found within the walls of the ventricles. They produce cerebrospinal fluid (CSF).

Occipital lobe

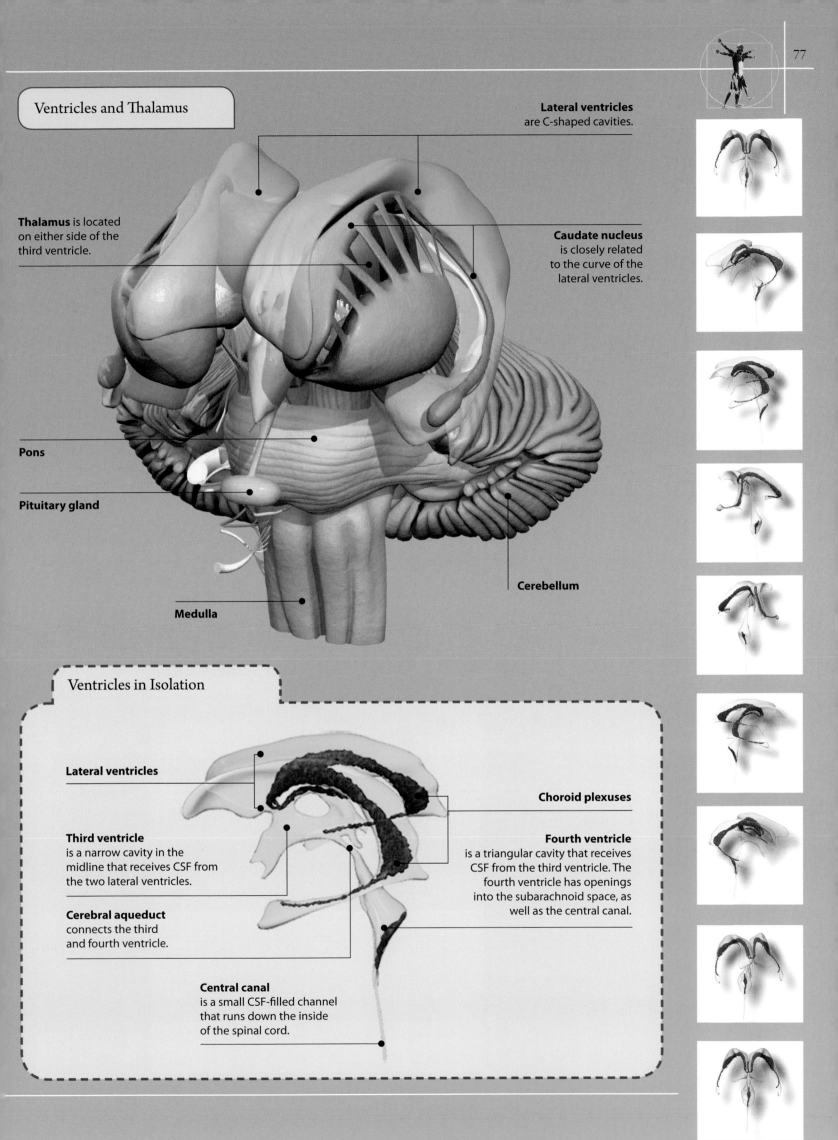

Ventricles and Thalamus

Lateral ventricles are C-shaped cavities.

Thalamus is located on either side of the third ventricle.

Caudate nucleus is closely related to the curve of the lateral ventricles.

Pons

Pituitary gland

Medulla

Cerebellum

Ventricles in Isolation

Lateral ventricles

Choroid plexuses

Third ventricle is a narrow cavity in the midline that receives CSF from the two lateral ventricles.

Fourth ventricle is a triangular cavity that receives CSF from the third ventricle. The fourth ventricle has openings into the subarachnoid space, as well as the central canal.

Cerebral aqueduct connects the third and fourth ventricle.

Central canal is a small CSF-filled channel that runs down the inside of the spinal cord.

The outer surface of the cerebrum, containing the nuclei of nerve cells, is called the cortex. It is tightly folded to increase the amount of nervous tissue that can be packed into the cranial cavity. It forms gyri (folds) and sulci (fissures between the folds). Some of these gyri and sulci are used to divide the brain up into different areas.

The cerebrum organizes many of the complex functions of the brain, including movement, touch, and pain sensation, speech, hearing, and vision. Different areas of the brain carry out these specific functions. Primary areas receive or send the nerve signals. Association areas help modify or interpret the nerve signals.

Central sulcus separates the frontal and parietal lobes.

Precentral gyrus is the gyrus immediately in front of the central sulcus. It controls motor function in the body.

Postcentral gyrus is the gyrus immediately behind the central sulcus. It receives sensory information from the body.

Longitudinal fissure runs in the midline from front to back, between the two cerebral hemispheres.

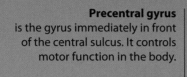

Lateral sulcus separates the temporal lobes from the frontal and parietal lobes.

Sulci are fissures between the gyri.

Gyri are folds of brain tissue.

Sulci and Gyri

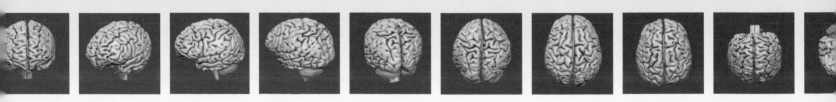

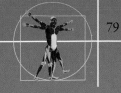

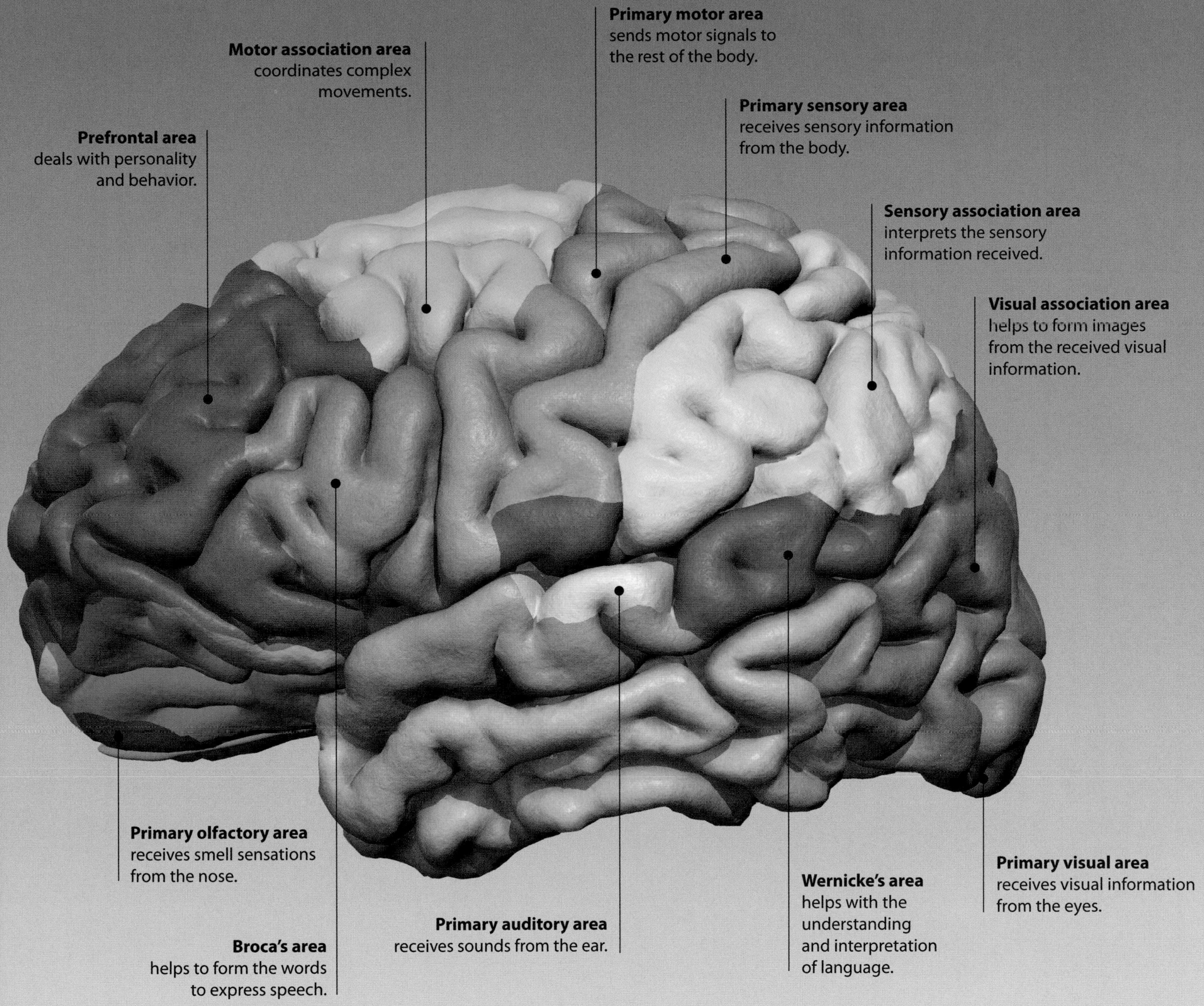

Motor association area
coordinates complex
movements.

Primary motor area
sends motor signals to
the rest of the body.

Primary sensory area
receives sensory information
from the body.

Prefrontal area
deals with personality
and behavior.

Sensory association area
interprets the sensory
information received.

Visual association area
helps to form images
from the received visual
information.

Primary olfactory area
receives smell sensations
from the nose.

Wernicke's area
helps with the
understanding
and interpretation
of language.

Primary visual area
receives visual information
from the eyes.

Broca's area
helps to form the words
to express speech.

Primary auditory area
receives sounds from the ear.

Cortical Areas of the Cerebrum

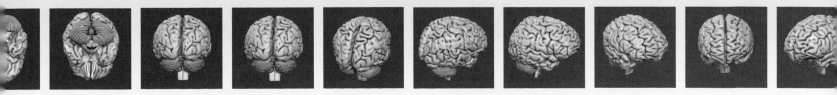

The primary motor and sensory areas of the brain are organized in such a way that each part of the body is represented by a specific region on these strips of cerebral cortex. The size of the cortex which represents the body part is a reflection of how many motor or sensory nerves supply that region.

    Homunculi (little humans) are representations of the entire human body. However, each body part is drawn on a scale corresponding to the size of the cortex and number of nerves which supply them, rather than their actual physical size. This creates rather distorted characters which give a graphic representation of the amount of cortex allocated to each body part. Both motor and sensory homunculi can be created.

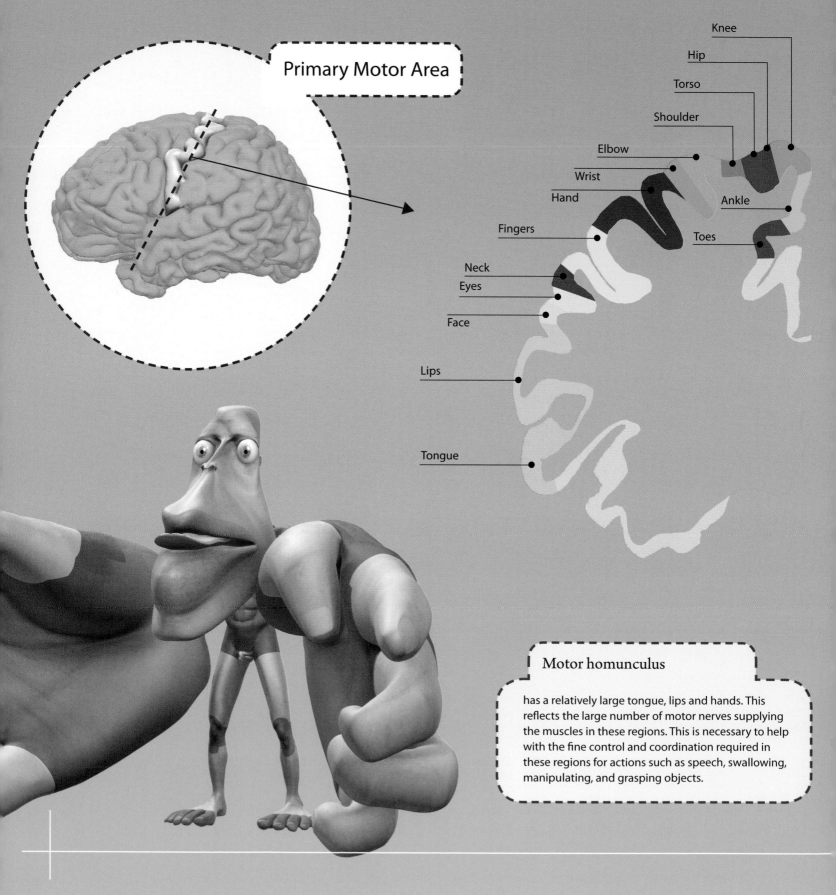

Primary Motor Area

Knee
Hip
Torso
Shoulder
Elbow
Wrist
Hand
Fingers
Neck
Eyes
Face
Lips
Tongue
Ankle
Toes

Motor homunculus

has a relatively large tongue, lips and hands. This reflects the large number of motor nerves supplying the muscles in these regions. This is necessary to help with the fine control and coordination required in these regions for actions such as speech, swallowing, manipulating, and grasping objects.

Primary Sensory Area

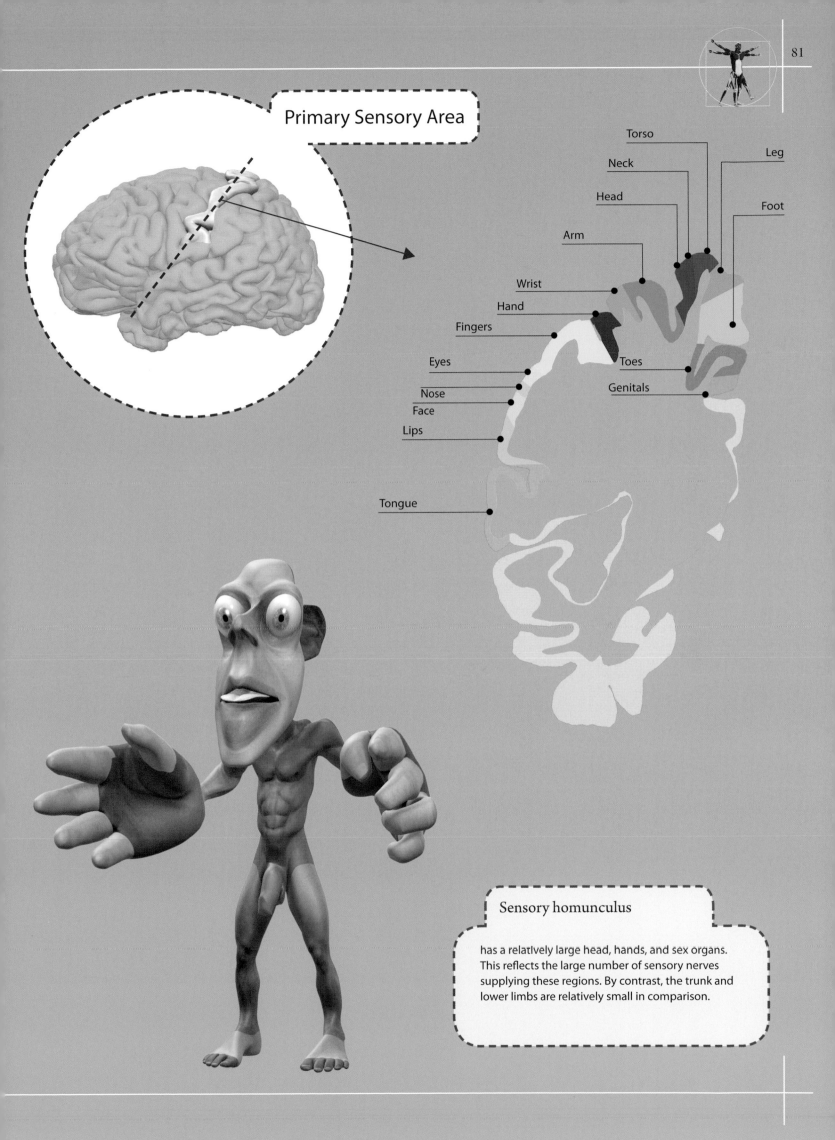

Torso

Neck

Head

Leg

Arm

Foot

Wrist

Hand

Fingers

Eyes

Toes

Nose

Genitals

Face

Lips

Tongue

Sensory homunculus

has a relatively large head, hands, and sex organs. This reflects the large number of sensory nerves supplying these regions. By contrast, the trunk and lower limbs are relatively small in comparison.

The basal ganglia are collections of nerve cell nuclei, located deep within the cerebrum. The three distinct areas are known as the caudate nucleus, putamen, and globus pallidus. Their main role is in coordinating and controlling body movements. They also influence muscle tone and body posture.

The thalamus is a large collection of nerve cell nuclei located between the brainstem and the cerebrum, close to the basal ganglia. It is a relay center for the many nerve fibers passing between the cerebral cortex (outer part of the cerebrum) and the brainstem. Connecting fibers running to the cerebral cortex form a band known as the internal capsule.

Did you know?

Parkinson's disease is a relatively common nervous system disease which affects the basal ganglia. It is caused by the loss of nerve fibers which release a substance called dopamine. Patients often have a tremor and have difficulty starting movements.

Internal capsule
is made up of bundles of nerve fibers connecting the thalamus and cerebral cortex.

Cerebral cortex (grey matter)
is the outer part of the cerebrum containing nerve cell bodies and nuclei.

White matter
is the deeper part of the cerebrum that contains nerve fibers.

Putamen
has connections to the caudate nucleus.

Corpus callosum
is a thick band of nerve fibers connecting the two cerebral hemispheres.

Caudate nucleus
forms a C-shaped structure, with a broad head and a tail.

Pituitary | **Pons** | **Medulla**

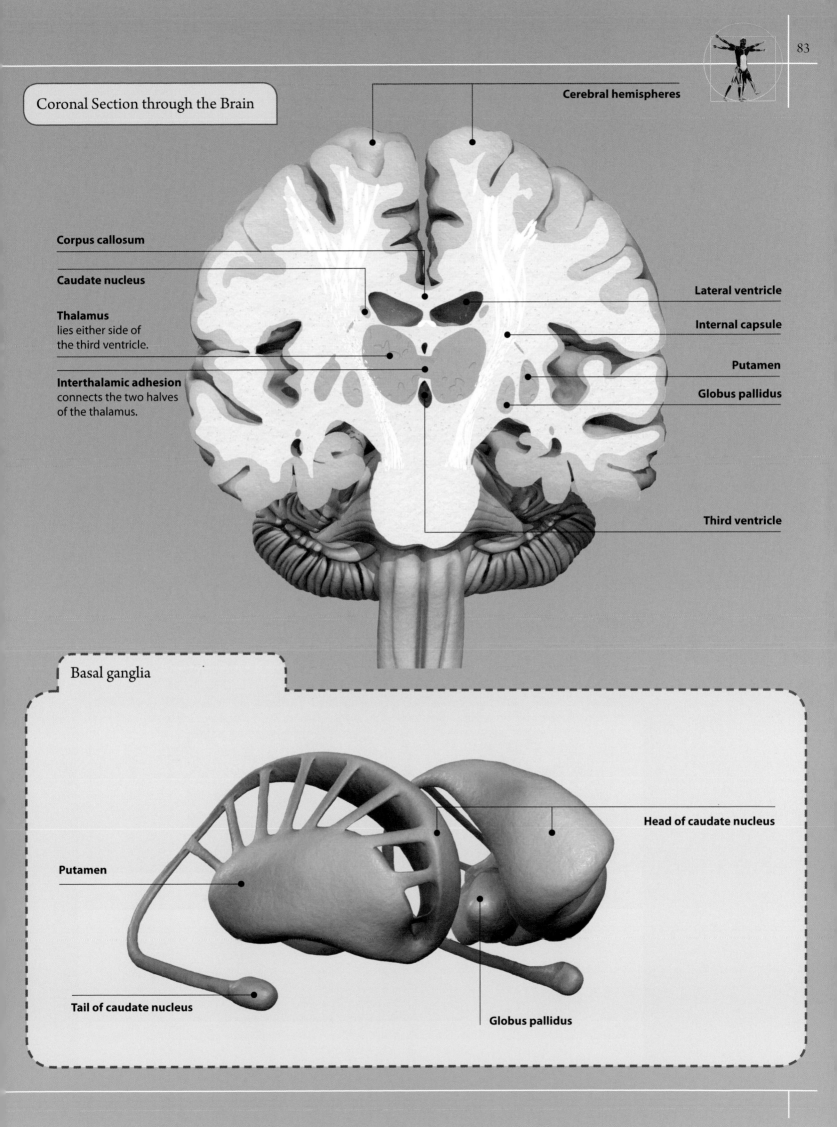

Coronal Section through the Brain

Cerebral hemispheres

Corpus callosum

Caudate nucleus

Thalamus
lies either side of
the third ventricle.

Interthalamic adhesion
connects the two halves
of the thalamus.

Lateral ventricle

Internal capsule

Putamen

Globus pallidus

Third ventricle

Basal ganglia

Head of caudate nucleus

Putamen

Tail of caudate nucleus

Globus pallidus

The brain requires a constant supply of blood. This is delivered by two paired vessels in the neck, the internal carotid, and vertebral arteries. Their branches supply the brainstem and cerebellum, before uniting at the base of the brain to form a ring of arteries called the "circle of Willis." From this, three pairs of cerebral arteries deliver blood to the entire cerebrum.

Unusually, the veins of the brain follow a completely different course to the arteries. Large sinuses within the dura mater collect venous blood from the cerebral veins, and eventually drain into the paired internal jugular veins.

Circle of Willis

Anterior cerebral arteries
supply the central parts of the brain.

Optic chiasm
is where the optic nerves meet and some of their fibers cross over.

Middle cerebral arteries
supply the sides of the brain.

Pituitary gland

Posterior cerebral arteries
supply the back of the brain.

Basilar artery
is formed from the two vertebral arteries. It runs up the front of the brainstem.

Internal carotid arteries
are large vessels that enter the skull from the neck. They form the anterior and middle cerebral arteries.

Cerebellar arteries
supply the cerebellum.

Vertebral arteries
are paired arteries that travel up the neck to supply the cerebellum, brainstem and back of the brain.

Veins of the Brain

Superior sagittal sinus
runs in the midline collecting blood from the cerebral veins.

Cerebral veins
drain blood from the cerebrum into the venous sinuses.

Transverse sinus
runs along the inside of the back of the skull, and drains into the sigmoid sinus.

Cavernous sinus
drains blood from the eye along with the front and side of the brain.

Sigmoid sinuses
are paired curved vessels that drain blood into the internal jugular veins.

Internal jugular veins
return blood from the brain back to the heart.

The eyes are the organs of vision. They are delicate, spherical structures, which, by focusing and converting light rays into nerve signals, allow us to see objects around us. Only a small part of the eyeball can be seen, as most of it is hidden and protected inside a bony cavity within the skull, called the orbit.

The lacrimal glands (tear glands) and eyelids provide lubrication and protection to the exposed front surface of the eye. Muscles attached to the outside of the eyeball move them in various directions.

Lateral commissure
is the meeting point of the upper and lower eyelids furthest away from the nose.

Lacrimal gland
produces tears that keep the exposed surface of the eye clean and moist.

Palpebral fissure
is the gap between the eyelids.

Bony orbit
almost completely surrounds the eyeball, providing protection and support.

Levator palpebrae superioris
is a muscle that opens the eye by raising the upper eyelid.

Eyeball
has a tough protective outer coat to which muscles can attach. Inside the eyeball are delicate structures that help focus and convert light rays into nerve signals.

Eyelids
contain muscles that help open and close the eyes.

Inferior oblique muscle
is one of the muscles that move the eyeball.

Iris
is the colored part at the front of the eye.

Pupil
is the round opening in the iris through which light can enter into the eyeball.

Orbit from above

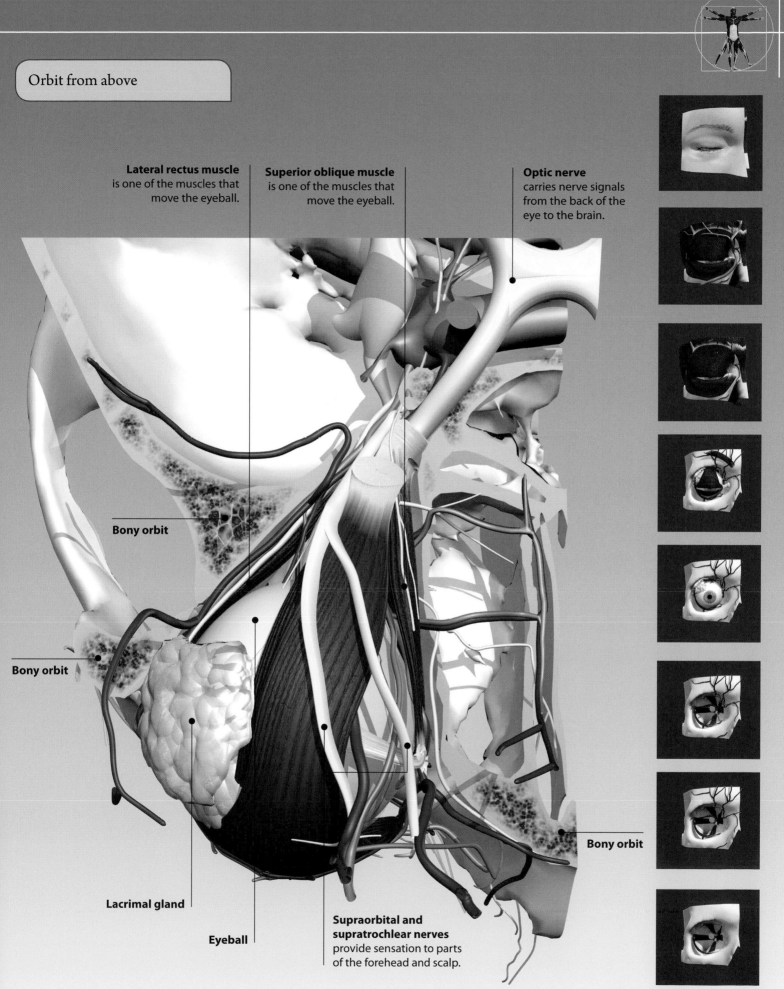

Lateral rectus muscle is one of the muscles that move the eyeball.

Superior oblique muscle is one of the muscles that move the eyeball.

Optic nerve carries nerve signals from the back of the eye to the brain.

Bony orbit

Bony orbit

Bony orbit

Lacrimal gland

Eyeball

Supraorbital and supratrochlear nerves provide sensation to parts of the forehead and scalp.

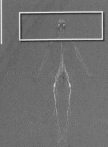

The eye is often compared to a camera. It has an opening at the front to vary the amount of light that enters it (iris), structures to focus the light rays (cornea and lens), and a light sensitive area (retina) to receive the focused image. It has a tough protective outer coat (sclera) and a "power supply" that provides its structures with blood and nutrients (choroid). There is even a "cable" (optic nerve) that "downloads" images formed on the retina directly to the part of the brain which interprets them.

Choroid
is a layer lying between the sclera and retina. It contains blood vessels that nourish the retina.

Pupil
is the round opening in the center of the iris. Muscle fibers in the iris can change the size of the pupil and control the amount of light entering the eye.

Cornea
This is the transparent curved surface covering the front of the eye. It helps to focus light rays entering the eye.

Iris
is the colored part of the eye.

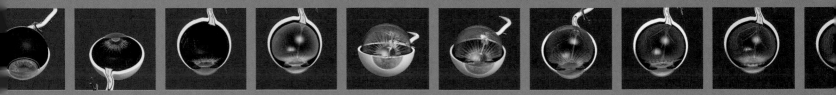

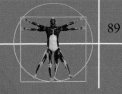

The Eye

Ciliary body
produces a watery fluid called aqueous humor, which circulates in the anterior cavity.

Posterior cavity
is the area behind the lens. It contains a thick jelly like material called the vitreous humor.

Retina
is a layer of specialized nerve cells lining the back and inside of the eye. It converts light rays into a signal that can be sent to the brain via the optic nerve.

Lens
is a clear flexible structure that is suspended behind the iris by suspensory ligaments attached to the ciliary body. It focuses light rays to form a clear image on the retina.

Sclera (or "white of the eye")
is the tough, protective outer layer that covers the entire eye, except the cornea.

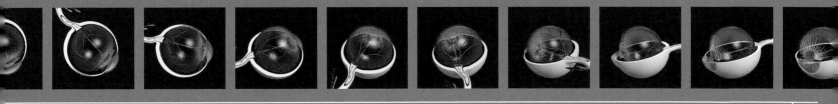

Vision is probably our most important sense. It allows us to identify different colored objects, over a range of distances, in both bright and dark light conditions.
The retina lines most of the inside of the eye. It converts light rays from an object into an electrical signal that can be sent to the brain for processing. The retina is made up of a number of layers, most of which are formed by different types of nerve cells.

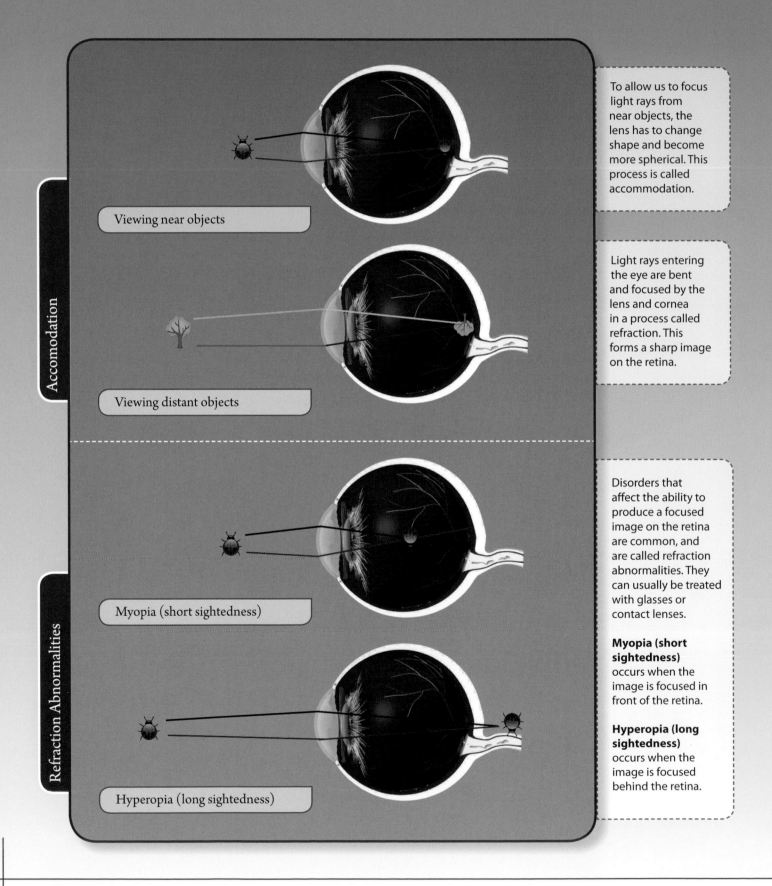

Accomodation

Viewing near objects

To allow us to focus light rays from near objects, the lens has to change shape and become more spherical. This process is called accommodation.

Viewing distant objects

Light rays entering the eye are bent and focused by the lens and cornea in a process called refraction. This forms a sharp image on the retina.

Refraction Abnormalities

Myopia (short sightedness)

Hyperopia (long sightedness)

Disorders that affect the ability to produce a focused image on the retina are common, and are called refraction abnormalities. They can usually be treated with glasses or contact lenses.

Myopia (short sightedness) occurs when the image is focused in front of the retina.

Hyperopia (long sightedness) occurs when the image is focused behind the retina.

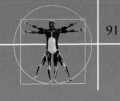
Microanatomy of the Retina

Bipolar cells
receive the electrical signals
from the photoreceptors.

Amacrine cells
and horizontal cells help 'fine tune'
the signal before it leaves the retina.

Ganglion cells
from all over the retina group
together and leave the eye as the
optic nerve. The signal is then
passed along the visual pathway
to those parts of the brain that
interpret visual information.

Structure of Rod and Cone Cells

There are two types of photoreceptors:
rods and cones. They are named according
to the shape of their outer segment.

Rods
help us to see in dim
light. Each retina has
about 120 million rods.

Cones
give us color vision.
There are about
6 million cones in
each retina.

Pigmented layer
is the outer layer of the retina. It
contains colored cells that absorb
any stray light rays.

LASIK—Laser-Assisted In-Situ Keratomileusis

LASIK is a simple surgical procedure
that can be used to correct refraction
abnormalities, without the need for
glasses or contact lenses. A special
laser reshapes the cornea so that light
rays are focused on the retina.

Photoreceptors
are the cells involved in
converting light rays to
electrical signals.

EXTRINSIC EYE MUSCLES

Six small muscles coordinate the rapid and precise movements of the eyeball. They allow us to follow objects through a wide range of motion, without us having to constantly move our head. They are all attached from the walls of the bony orbit to the tough white outer part of the eyeball, called the sclera. They get their nerve supply from cranial nerves III, IV, and VI.

Optic nerve
leaves the back of the eyeball and carries visual information to the brain.

Superior rectus
attaches to the top of the eyeball.

Trochlea
is a small loop of fibrous tissue at the top of the bony orbit, which acts like a pulley for the tendon of the superior oblique muscle.

Inferior oblique

Superior rectus

Lateral rectus

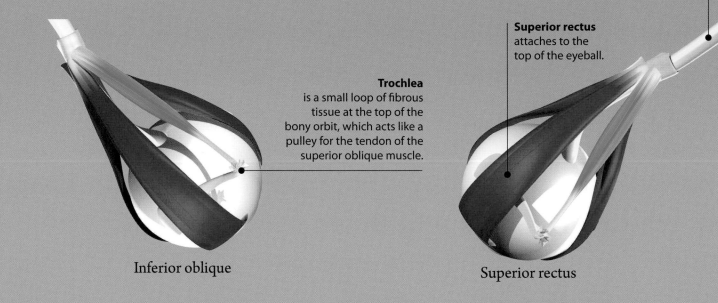

Medial rectus

Superior oblique

Inferior rectus
attaches to the bottom of the eyeball.

Inferior rectus

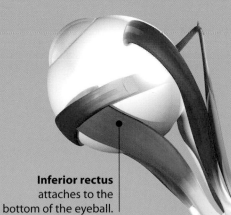

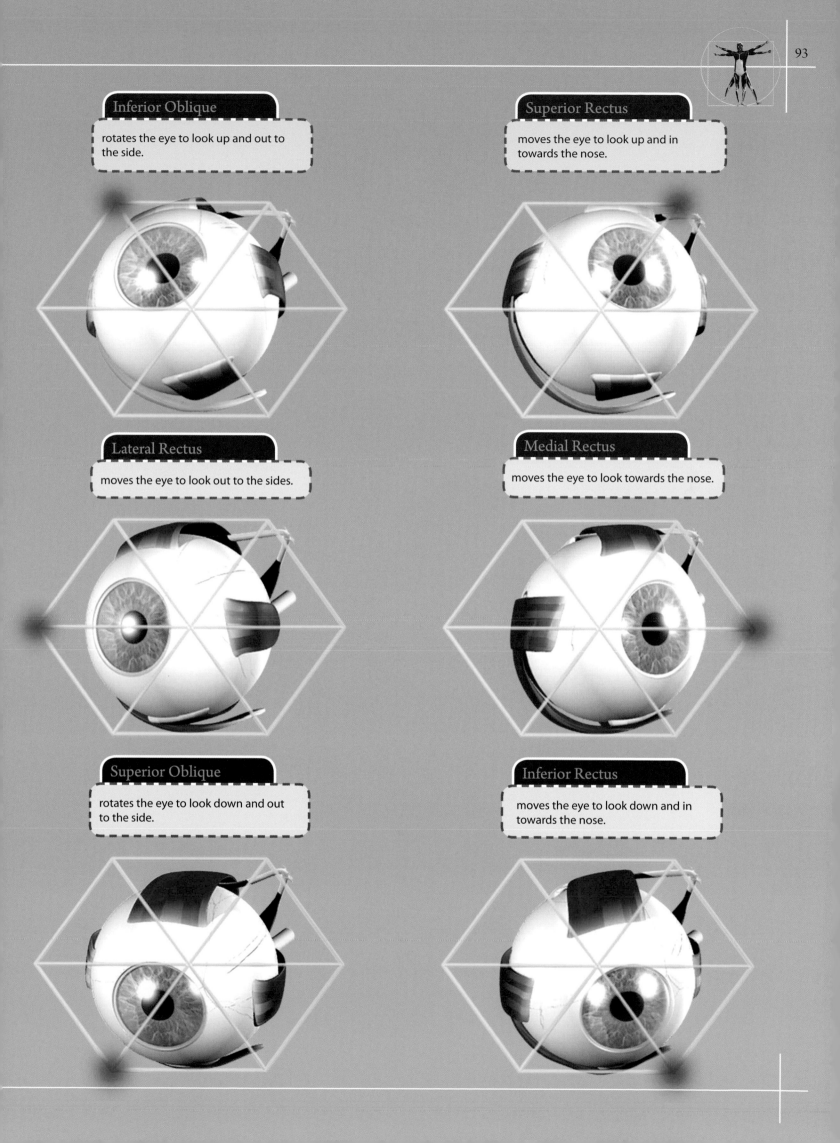

Inferior Oblique

rotates the eye to look up and out to the side.

Superior Rectus

moves the eye to look up and in towards the nose.

Lateral Rectus

moves the eye to look out to the sides.

Medial Rectus

moves the eye to look towards the nose.

Superior Oblique

rotates the eye to look down and out to the side.

Inferior Rectus

moves the eye to look down and in towards the nose.

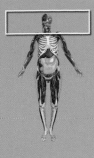

The lacrimal apparatus is a collection of structures involved with the production and drainage of tears. Tears are made by the lacrimal glands, and help moisturize and protect the surface of the eye. When we blink, or close our eyes, tears are swept across the surface of the eye towards the nose. The tears then drain via small openings on our eyelids (puncta), into a channel that leads eventually into the nose (nasolacrimal duct). This explains why, when we cry, our nose often runs as well.

Did you know?

The average rate of blinking is about 10 times per minute in adults. This reduces to about 3-4 times per minute when reading or concentrating on a screen, and can lead to sore eyes, as they are less moisturized and start to dry out.

Eyeball

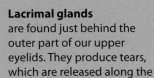

Lacrimal glands
are found just behind the outer part of our upper eyelids. They produce tears, which are released along the edge of the upper eyelid.

Tarsal muscles
form part of the eyelids and assist with opening the eyes.

Lacrimal canals
carry tears from the puncta to the lacrimal sac.

Lacrimal sac
is the wide upper part of the nasolacrimal duct.

Nasolacrimal duct
drains tears into the nasal cavity.

Lacrimal puncta
are small openings found on the part of the eyelids closest to the nose. They drain tears from the surface of the eye.

Nasal cavity
is the space found inside the nose.

Skull

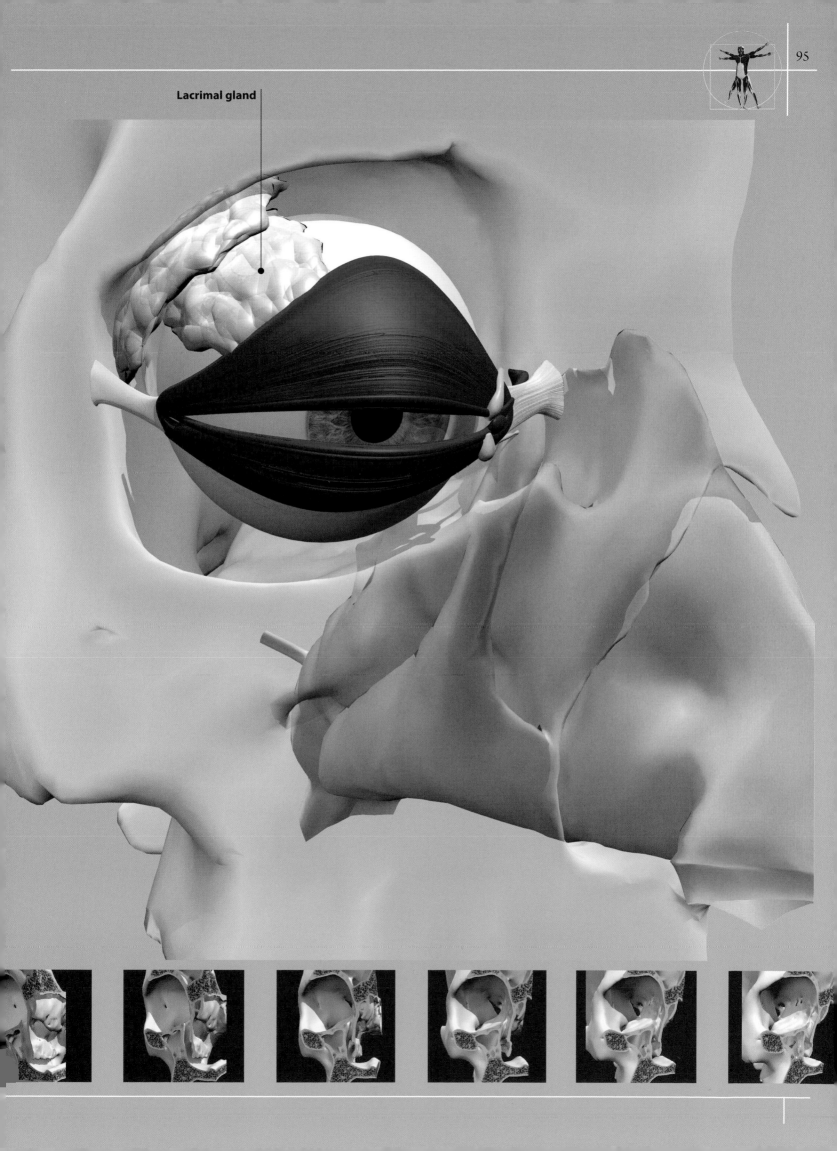

Lacrimal gland

PARANASAL SINUSES

The paranasal sinuses are air-filled cavities, found within the bones of the skull around the nose. There are four groups of sinuses: the maxillary, frontal, ethmoid, and sphenoid. They are connected to the nasal cavity, and lined by the same cells, producing a mucuslike secretion. This helps to moisten the air being breathed in, as well as trapping any particles. The main function of the paranasal sinuses is to reduce the weight of the skull. In addition, they also act like echo chambers and contribute to the sound of the voice.

Paranasal Sinuses

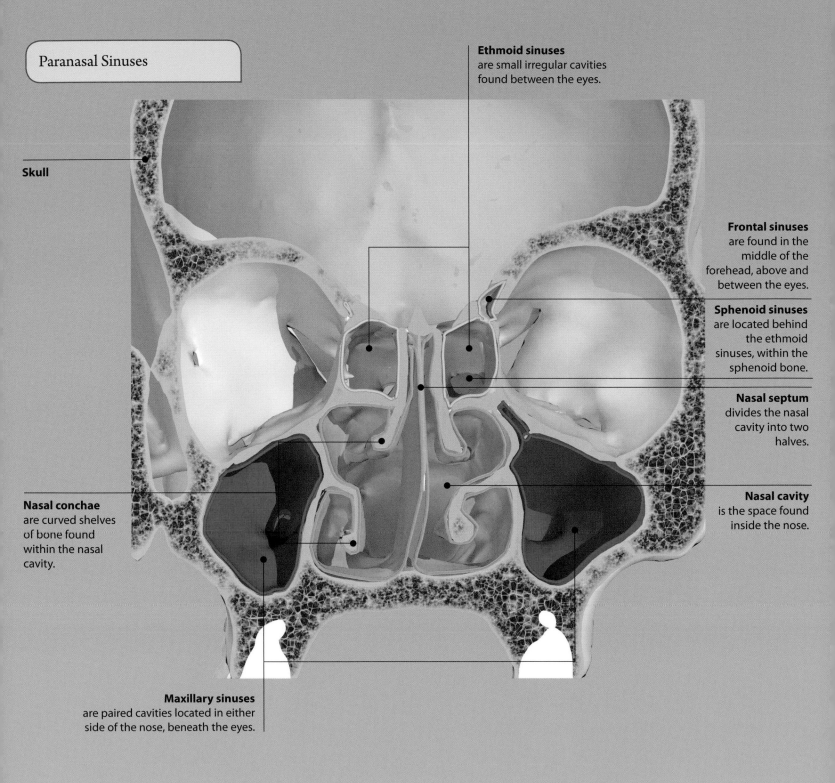

Ethmoid sinuses are small irregular cavities found between the eyes.

Skull

Frontal sinuses are found in the middle of the forehead, above and between the eyes.

Sphenoid sinuses are located behind the ethmoid sinuses, within the sphenoid bone.

Nasal septum divides the nasal cavity into two halves.

Nasal cavity is the space found inside the nose.

Nasal conchae are curved shelves of bone found within the nasal cavity.

Maxillary sinuses are paired cavities located in either side of the nose, beneath the eyes.

Coronal Section

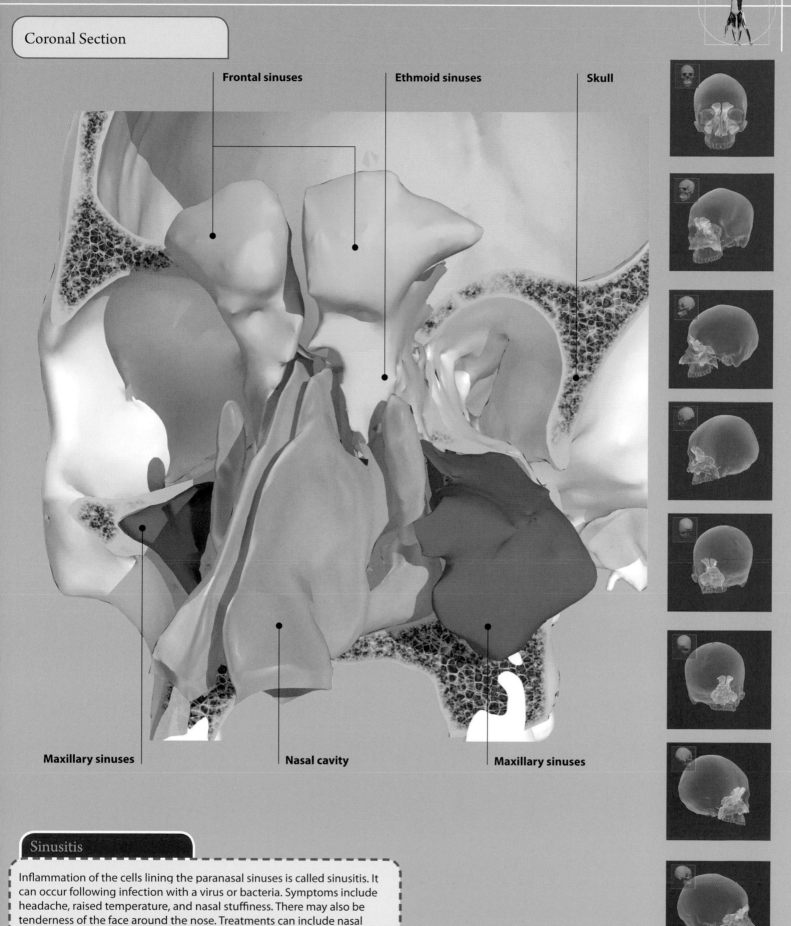

Frontal sinuses

Ethmoid sinuses

Skull

Maxillary sinuses

Nasal cavity

Maxillary sinuses

Sinusitis

Inflammation of the cells lining the paranasal sinuses is called sinusitis. It can occur following infection with a virus or bacteria. Symptoms include headache, raised temperature, and nasal stuffiness. There may also be tenderness of the face around the nose. Treatments can include nasal sprays, decongestants, pain killers, and sometimes antibiotics.

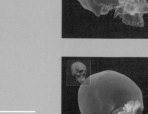

The ear is the organ of hearing and balance. Most of it is hidden from view, protected within the temporal bone of the skull. The ear is divided into three regions: outer, middle, and inner.

The hearing part of the ear detects sound waves over a wide range of frequencies and volumes, and sends nerve signals to the auditory parts of the brain to be interpreted. The balance (or vestibular) part of the brain detects movement and position of the head.

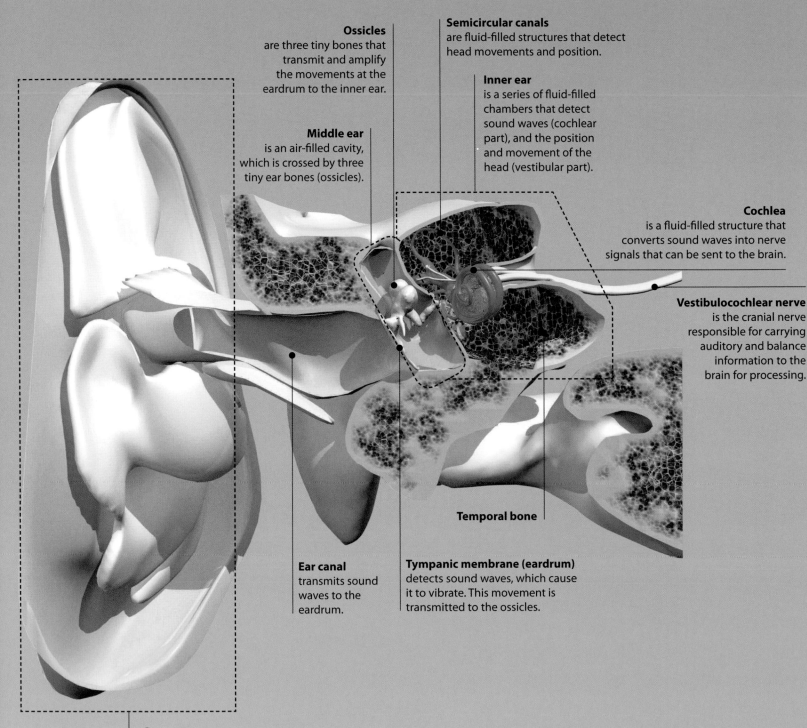

Ossicles
are three tiny bones that transmit and amplify the movements at the eardrum to the inner ear.

Semicircular canals
are fluid-filled structures that detect head movements and position.

Inner ear
is a series of fluid-filled chambers that detect sound waves (cochlear part), and the position and movement of the head (vestibular part).

Middle ear
is an air-filled cavity, which is crossed by three tiny ear bones (ossicles).

Cochlea
is a fluid-filled structure that converts sound waves into nerve signals that can be sent to the brain.

Vestibulocochlear nerve
is the cranial nerve responsible for carrying auditory and balance information to the brain for processing.

Temporal bone

Ear canal
transmits sound waves to the eardrum.

Tympanic membrane (eardrum)
detects sound waves, which cause it to vibrate. This movement is transmitted to the ossicles.

Outer ear
consists of the auricle and the ear canal. It is separated from the middle ear by the eardrum (tympanic membrane).

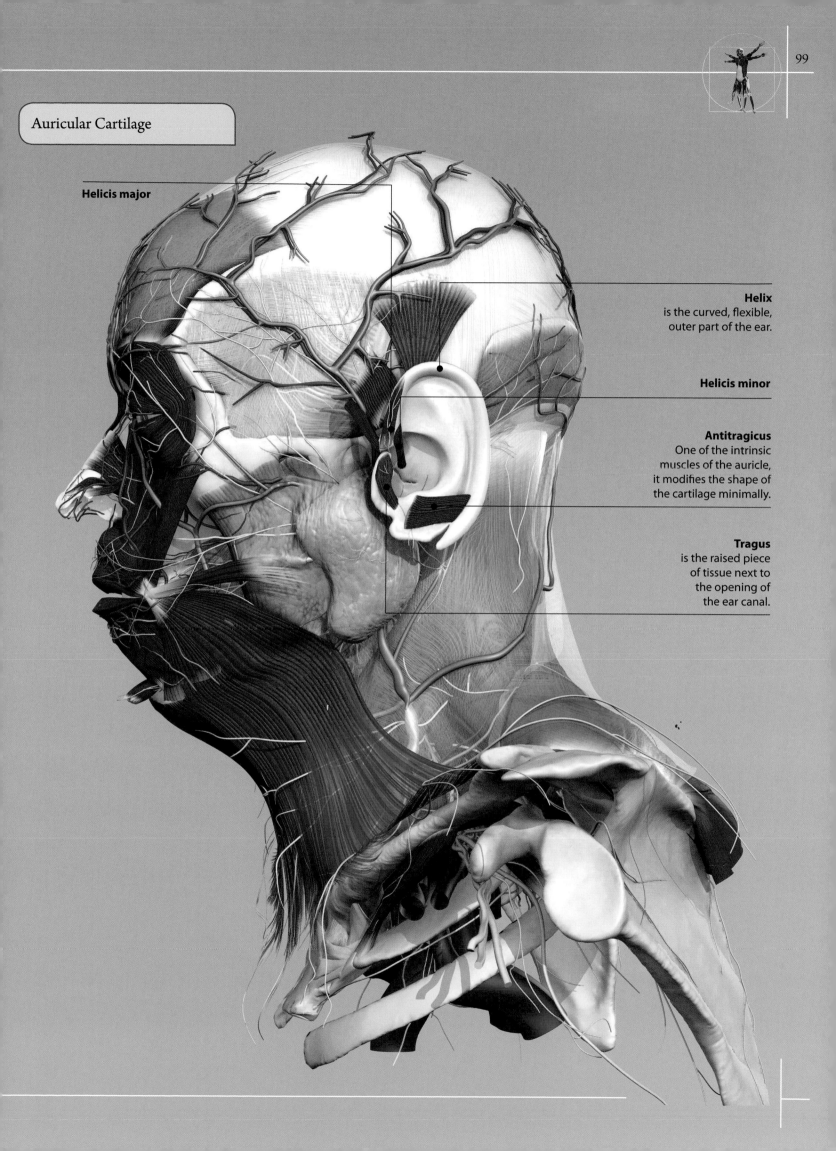

Auricular Cartilage

Helicis major

Helix
is the curved, flexible,
outer part of the ear.

Helicis minor

Antitragicus
One of the intrinsic
muscles of the auricle,
it modifies the shape of
the cartilage minimally.

Tragus
is the raised piece
of tissue next to
the opening of
the ear canal.

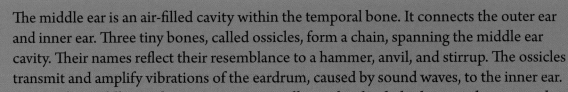

The middle ear is an air-filled cavity within the temporal bone. It connects the outer ear and inner ear. Three tiny bones, called ossicles, form a chain, spanning the middle ear cavity. Their names reflect their resemblance to a hammer, anvil, and stirrup. The ossicles transmit and amplify vibrations of the eardrum, caused by sound waves, to the inner ear.

The middle ear also contains two small muscles that help dampen down sound waves from excessively loud noises. The middle ear cavity is connected to the back of the throat via the Eustachian tube.

Stapes (stirrup)
is the final part of the ossicle chain. It transmits any vibrations to the inner ear, via the oval window.

Inner ear
contains the cochlea.

Stapedius muscle
attaches to the stapes bone, and helps dampen loud noises.

Vestibulocochlear nerve
carries the auditory nerve signals to the brain.

Incus (anvil)
is the middle of the three ossicles, lying between the malleus and the stapes.

Malleus (hammer)
is directly attached to the tympanic membrane, and is able to pass on any vibrations to the incus.

Middle ear cavity

Tympanic membrane (eardrum)
is a thin, semi-translucent piece of tissue stretched across the ear canal. It separates the outer ear from the middle ear. It vibrates in response to sound waves, and these movements are passed on to the ossicles.

Temporal bone

Tensor tympani
is a small muscle that tenses the tympanic membrane when exposed to sudden loud noises, reducing the volume of these noises.

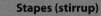

Ossicles in situ

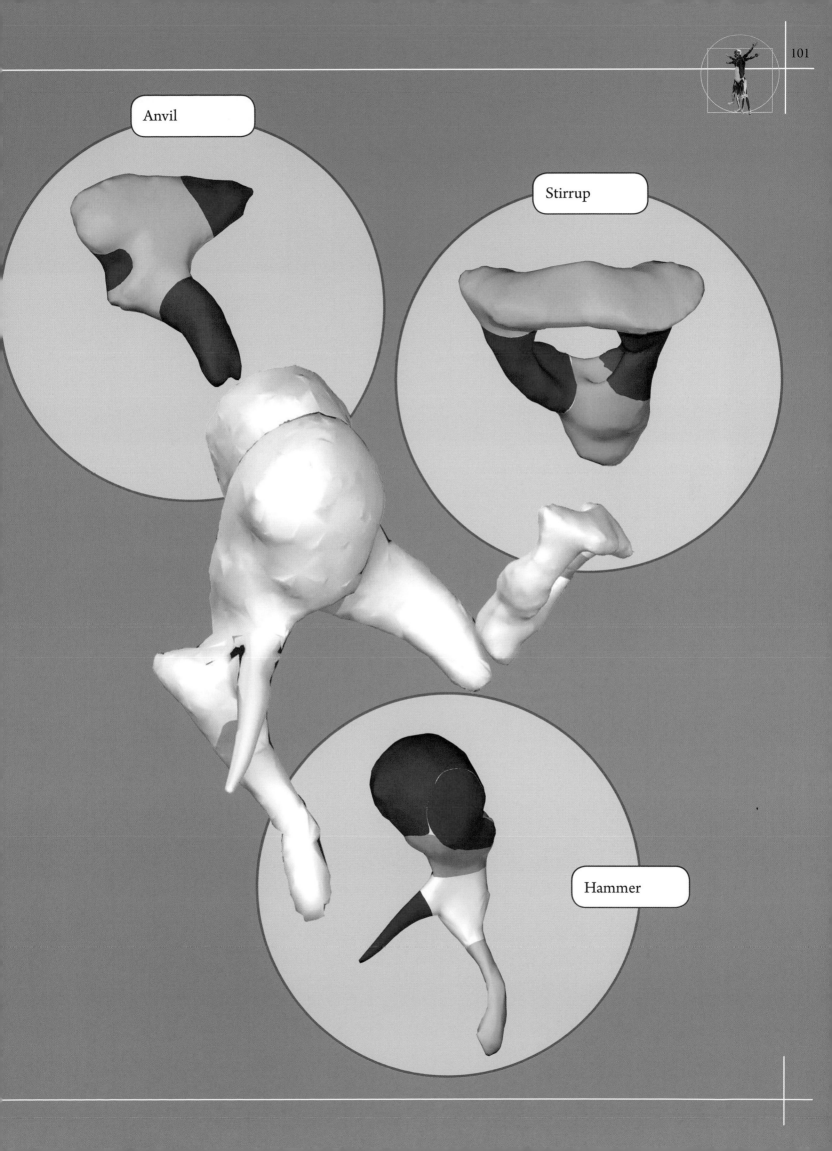

Anvil

Stirrup

Hammer

The inner ear is also known as the labyrinth. It has both bony and membranous parts. The bony labyrinth is found deep within the temporal bone of the skull. It consists of a series of complex, fluid-filled channels that are involved with hearing (cochlea) and balance (vestibule and semicircular canals).

Inside the bony labyrinth is found the membranous labyrinth. It also contains many fluid-filled channels. These often follow the shape of the bony labyrinth, and structures within them contribute to hearing and balance.

Membranous Labyrinth

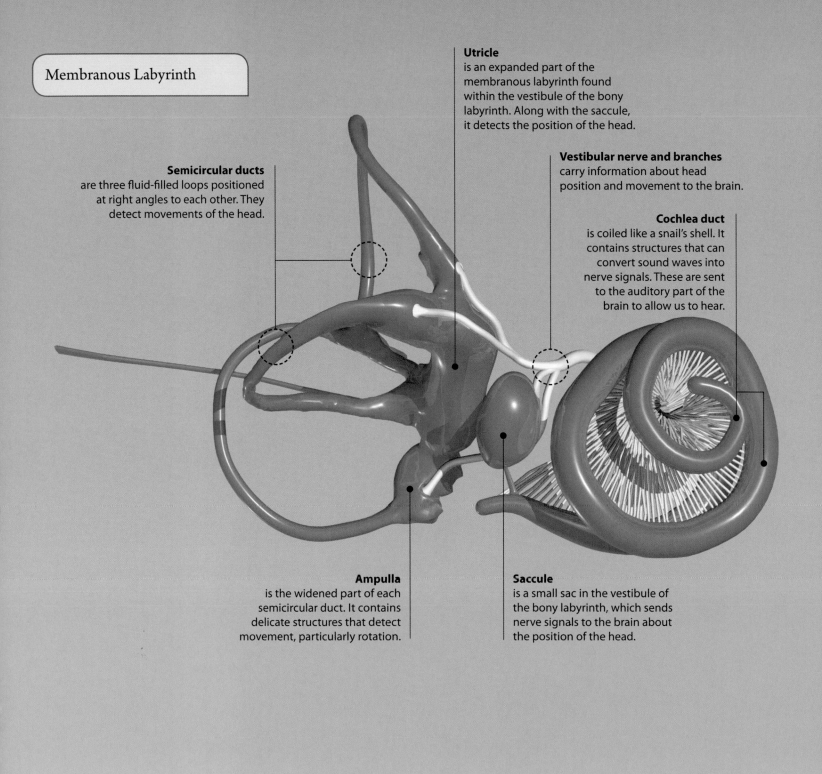

Utricle
is an expanded part of the membranous labyrinth found within the vestibule of the bony labyrinth. Along with the saccule, it detects the position of the head.

Vestibular nerve and branches
carry information about head position and movement to the brain.

Semicircular ducts
are three fluid-filled loops positioned at right angles to each other. They detect movements of the head.

Cochlea duct
is coiled like a snail's shell. It contains structures that can convert sound waves into nerve signals. These are sent to the auditory part of the brain to allow us to hear.

Ampulla
is the widened part of each semicircular duct. It contains delicate structures that detect movement, particularly rotation.

Saccule
is a small sac in the vestibule of the bony labyrinth, which sends nerve signals to the brain about the position of the head.

Inner Ear

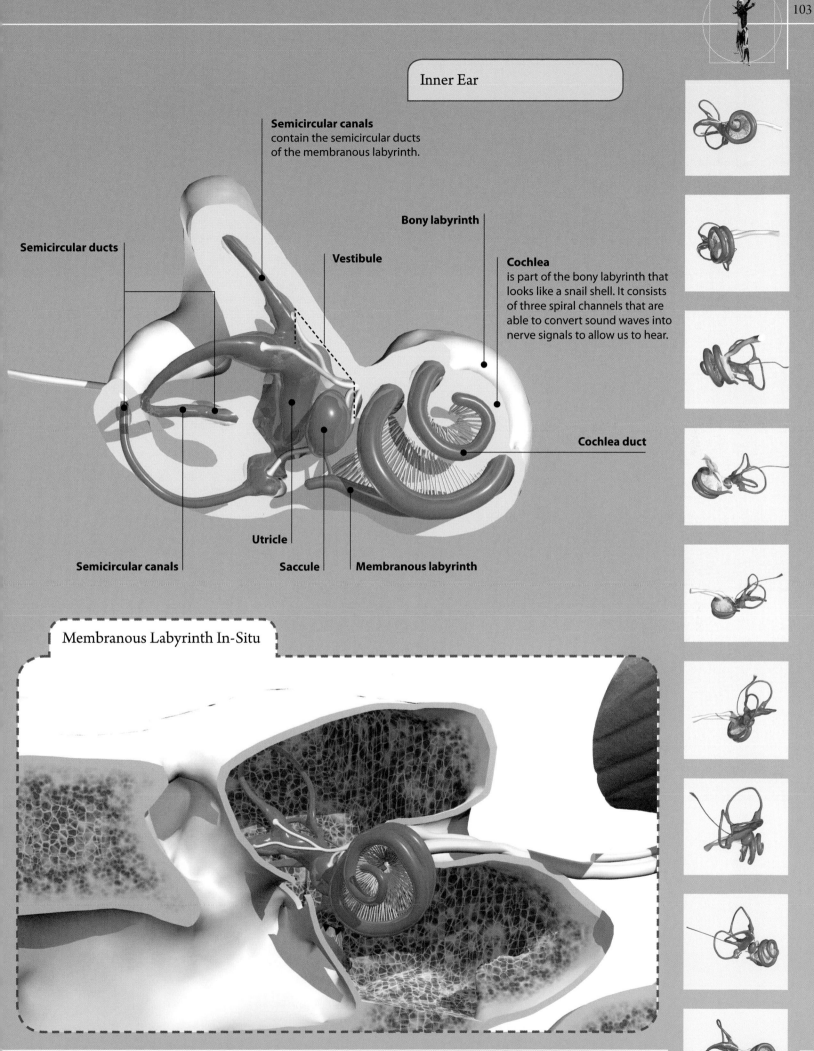

Semicircular canals
contain the semicircular ducts
of the membranous labyrinth.

Bony labyrinth

Semicircular ducts

Vestibule

Cochlea
is part of the bony labyrinth that
looks like a snail shell. It consists
of three spiral channels that are
able to convert sound waves into
nerve signals to allow us to hear.

Cochlea duct

Utricle

Semicircular canals

Saccule

Membranous labyrinth

Membranous Labyrinth In-Situ

Sound waves transmitted from the middle ear to the inner ear generate pressure waves in three fluid-filled spiral channels in the cochlea. A collection of specialized lining cells and associated structures, called the organ of Corti, detect these pressure changes. They convert them into nerve signals that can be sent to the brain and allow us to hear. The exact region of the cochlea stimulated depends upon the pitch of the sound, and this allows us to interpret different noises.

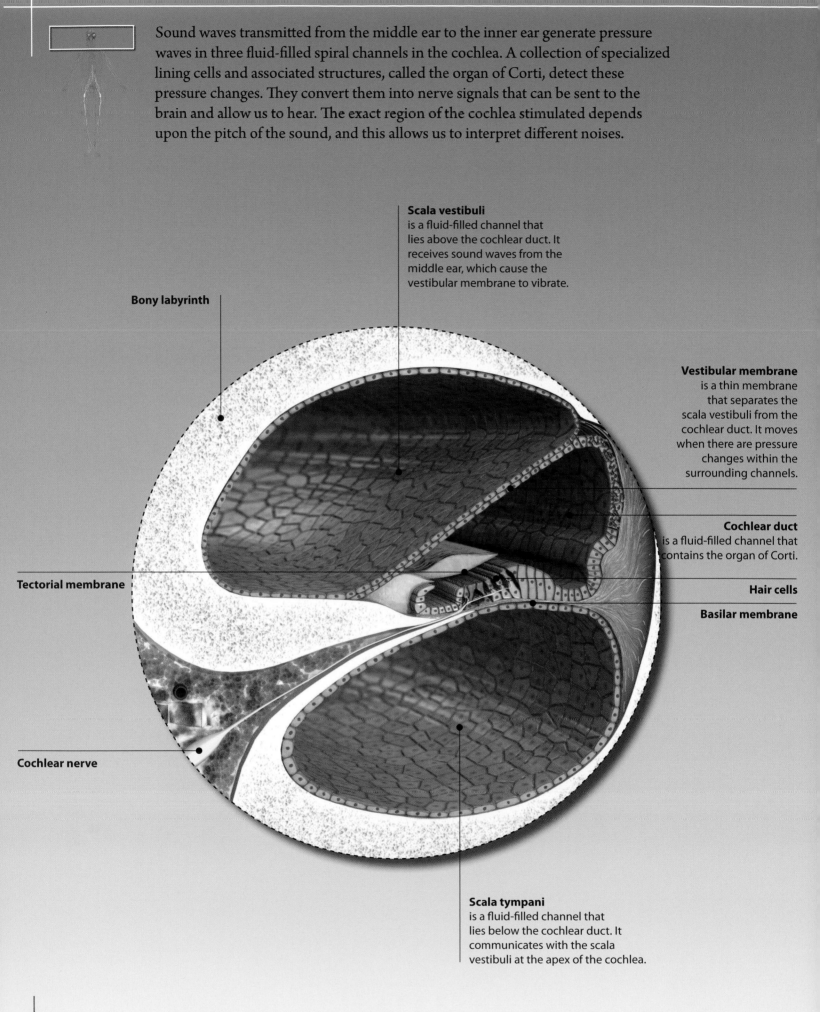

Scala vestibuli
is a fluid-filled channel that lies above the cochlear duct. It receives sound waves from the middle ear, which cause the vestibular membrane to vibrate.

Bony labyrinth

Vestibular membrane
is a thin membrane that separates the scala vestibuli from the cochlear duct. It moves when there are pressure changes within the surrounding channels.

Cochlear duct
is a fluid-filled channel that contains the organ of Corti.

Tectorial membrane

Hair cells

Basilar membrane

Cochlear nerve

Scala tympani
is a fluid-filled channel that lies below the cochlear duct. It communicates with the scala vestibuli at the apex of the cochlea.

Tectorial membrane
lies over the hair cells. It presses against and bends the stereocilia if the basilar membrane moves up in response to sound waves.

Vestibular membrane

Cochlear duct

Stereocilia

Hair cells
have specialized long surface projections called stereocilia. These can be bent by the overlying tectorial membrane, opening membrane ion channels and generating a nerve signal.

Cochlear nerve
carries nerve signals from the hair cells to the auditory part of the brain which allows us to hear.

Basilar membrane
lies beneath the organ of Corti. Movement of this layer in response to pressure waves, presses the hair cells against the tectorial membrane, causing the stereocilia to bend. This sends nerve signals to the brain.

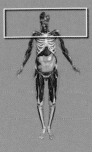

The upper aerodigestive tract includes organs such as the nose, mouth, throat, and larynx (voice box). They form connected passageways through which air, food, and water can enter the body, and also provide us with the special senses of taste and smell. Some parts of the upper aerodigestive tract are parts of both the respiratory and digestive systems.

Soft palate
helps prevent food and drink entering the nasal cavity.

Pharynx (throat)
can be divided into three regions: nasopharynx, oropharynx, and laryngopharynx.

Nasal cavity
is the large air-filled space inside the nose that humidifies, filters, and warms inhaled air. Specialized nerves at the top of the cavity allow us to smell.

Nose
is formed from bone and cartilage. It has two openings into the nasal cavity.

Hard palate
is a bony partition that separates the oral cavity from the nasal cavity.

Lips
cover the entrance to the oral cavity.

Teeth
allow us to cut and grind food into smaller pieces, to aid with digestion.

Mandible
is the lower jaw bone.

Tongue
is a muscular organ that helps with speech and swallowing. It also provides our sense of taste.

Oral cavity (mouth)
contains the teeth and tongue.

Larynx (voice box)
leads into the trachea (windpipe) and lungs. It contains the vocal cords, which vibrate and produce sounds when air passes by them, allowing us to speak.

Upper Aerodigestive Tract

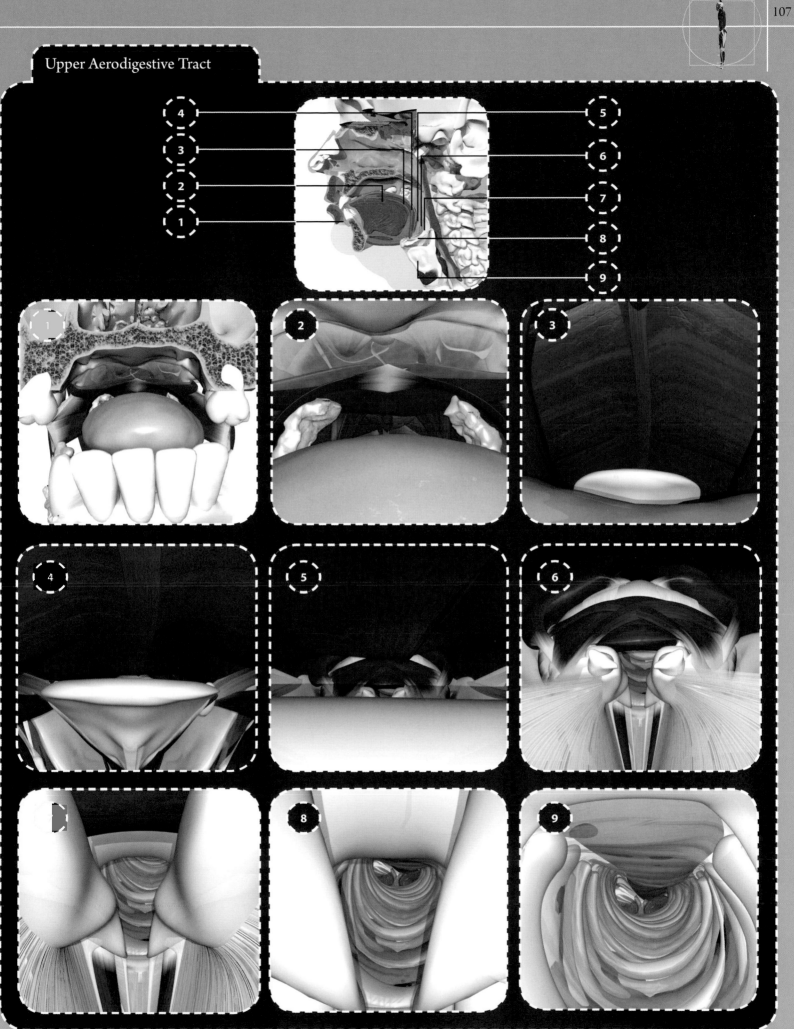

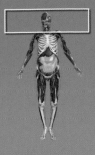

The nose is our organ of smell. It forms part of the respiratory system, and has an important role in cleaning and moistening the air we breathe in.

The nose has an outer part and an inner part. The outer part is the visible, pyramid-shaped structure in the middle of our faces. It is made of bone, cartilage, and fibrofatty tissue. There are two openings, directed downwards, on the outer nose. These lead into the inner part of the nose, the nasal cavity, where the specialized smell receptors are found.

The nose has a rich blood supply, which helps warm and moisten incoming air, but also means that it can bleed easily if it is damaged.

Procerus
is a muscle that attaches to the bridge of the nose. It wrinkles the nose.

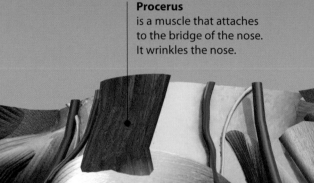

Nasalis (transverse part)
is a muscle that passes over the nose, and acts to narrow the nostrils.

Nasalis (alar parts)
are muscles attached around the openings of the nose. They widen, and flare the nostrils.

Lateral nasal artery
is a branch of the facial artery. It supplies the sides of the nose.

Fibrofatty tissue
gives shape and support to the nose, along with the bone and cartilage.

Apex
is the tip of the nose.

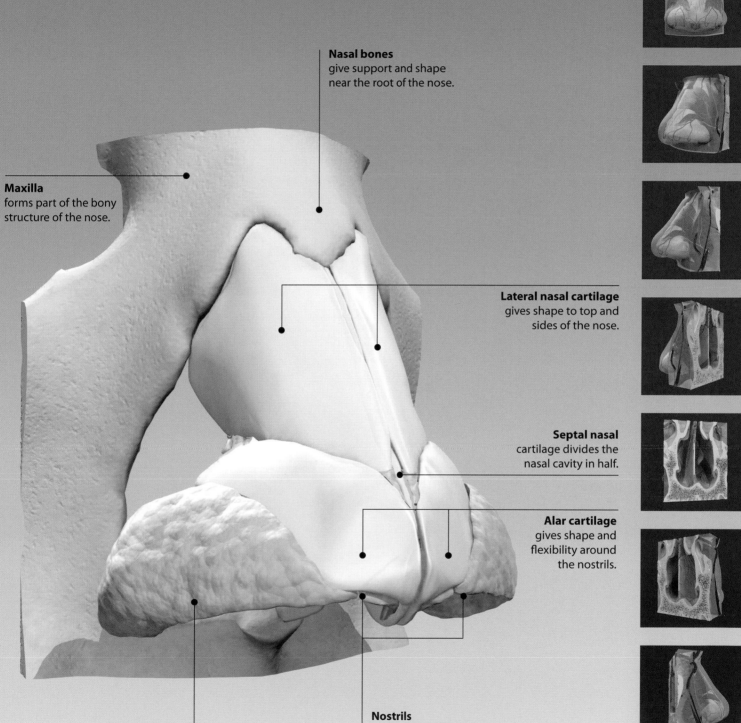

Nasal bones
give support and shape
near the root of the nose.

Maxilla
forms part of the bony
structure of the nose.

Lateral nasal cartilage
gives shape to top and
sides of the nose.

Septal nasal
cartilage divides the
nasal cavity in half.

Alar cartilage
gives shape and
flexibility around
the nostrils.

Nostrils
are the two openings of the
external nose. They lead
into the nasal cavity.

Fibrofatty tissue
provides support around the nostrils.

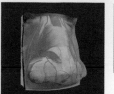

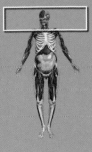

The nasal cavity is the inner part of the nose. It is the space within the skull between the nostrils and the nasopharynx, and is split in half by a vertical bony partition called the septum. The nasal cavity warms, filters, and moistens the air that is breathed in. The nerves that detect smell are found at the top of the nasal cavity.

Brain
processes the nerve signals sent from the olfactory nerves to allow us to recognize different smells.

Olfactory tract
delivers nerve signals related to smell to the brain.

Olfactory bulb
is a swelling at the end of the olfactory tract. It receives the olfactory nerve signals and passes them along the olfactory tract to the brain.

Olfactory nerves (CN I)
are the first cranial nerves. They carry nerve signals from odor receptors to the olfactory bulb.

Olfactory epithelium
is found at the top of the nasal cavity. It contains the specialized olfactory receptor cells that detect different odors.

Nasal cavity

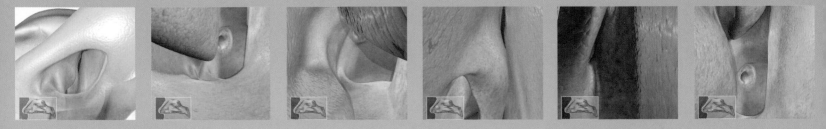

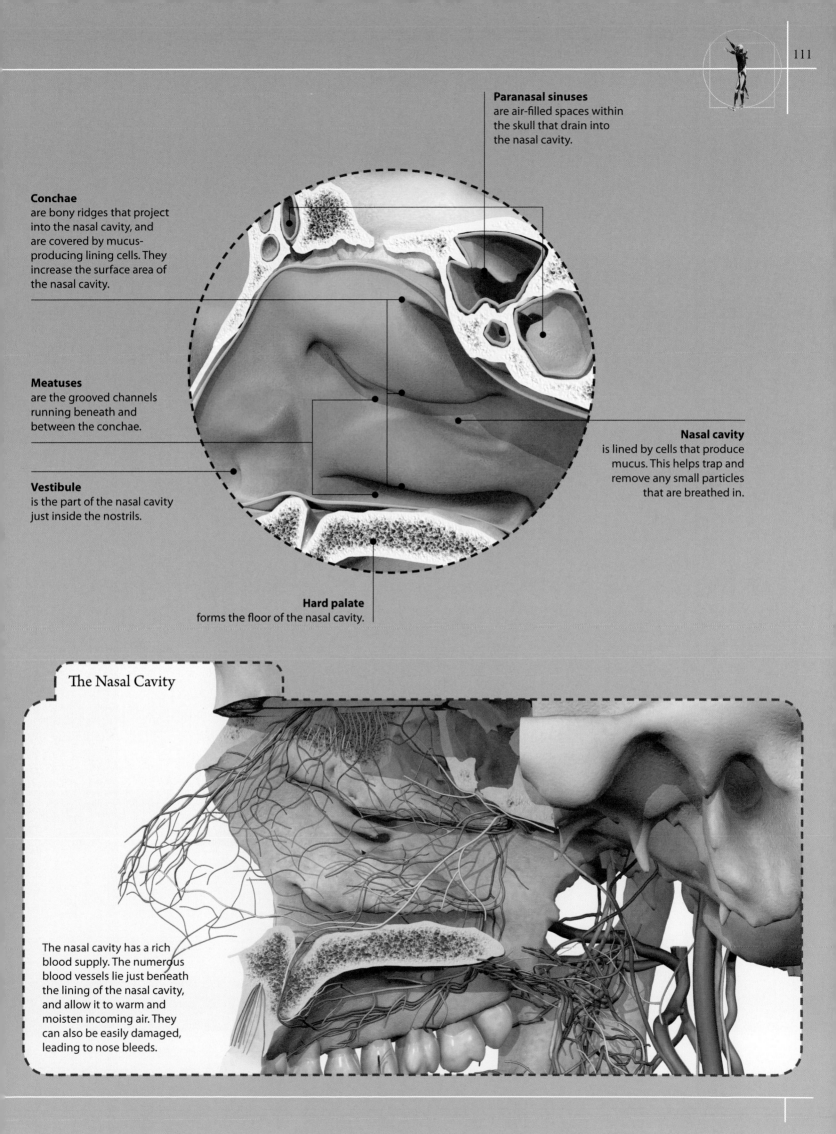

Paranasal sinuses
are air-filled spaces within the skull that drain into the nasal cavity.

Conchae
are bony ridges that project into the nasal cavity, and are covered by mucus-producing lining cells. They increase the surface area of the nasal cavity.

Meatuses
are the grooved channels running beneath and between the conchae.

Vestibule
is the part of the nasal cavity just inside the nostrils.

Nasal cavity
is lined by cells that produce mucus. This helps trap and remove any small particles that are breathed in.

Hard palate
forms the floor of the nasal cavity.

The Nasal Cavity

The nasal cavity has a rich blood supply. The numerous blood vessels lie just beneath the lining of the nasal cavity, and allow it to warm and moisten incoming air. They can also be easily damaged, leading to nose bleeds.

Smell, or olfaction, is one of the five special senses. Humans are able to distinguish thousands of different smells.

The olfactory epithelium at the top of the nasal cavity contains millions of olfactory receptor cells. These specialized nerve cells detect different odor causing molecules (odorants) that have been breathed in and dissolved in the mucus coating the olfactory epithelium. These odorants generate a nerve signal that is passed to the brain for processing and the recognition of a distinct smell.

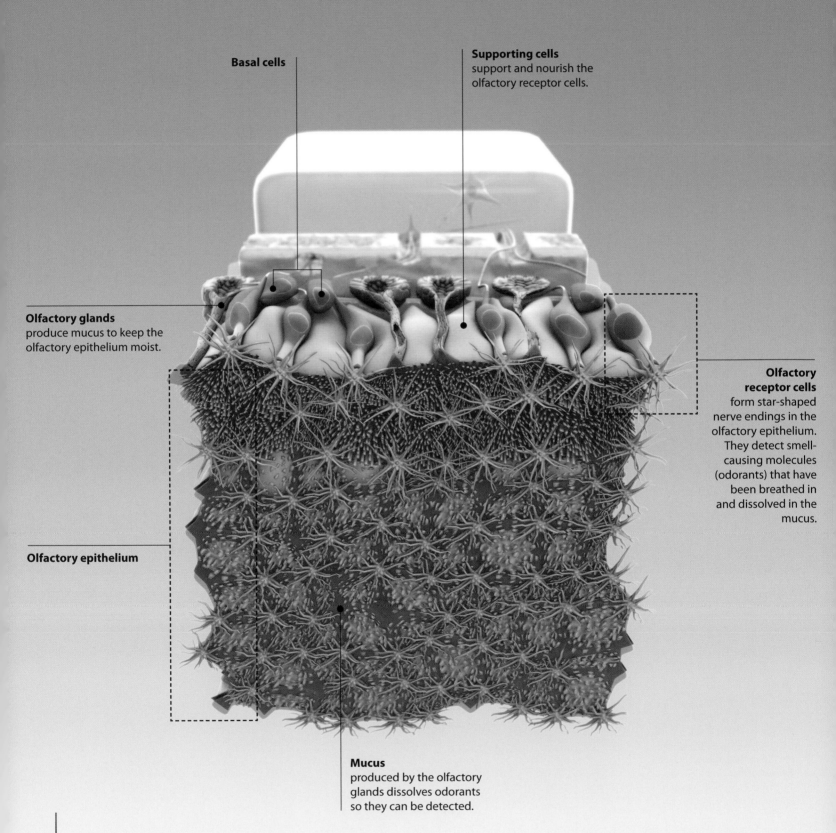

Basal cells

Supporting cells
support and nourish the olfactory receptor cells.

Olfactory glands
produce mucus to keep the olfactory epithelium moist.

Olfactory receptor cells
form star-shaped nerve endings in the olfactory epithelium. They detect smell-causing molecules (odorants) that have been breathed in and dissolved in the mucus.

Olfactory epithelium

Mucus
produced by the olfactory glands dissolves odorants so they can be detected.

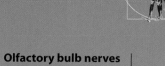

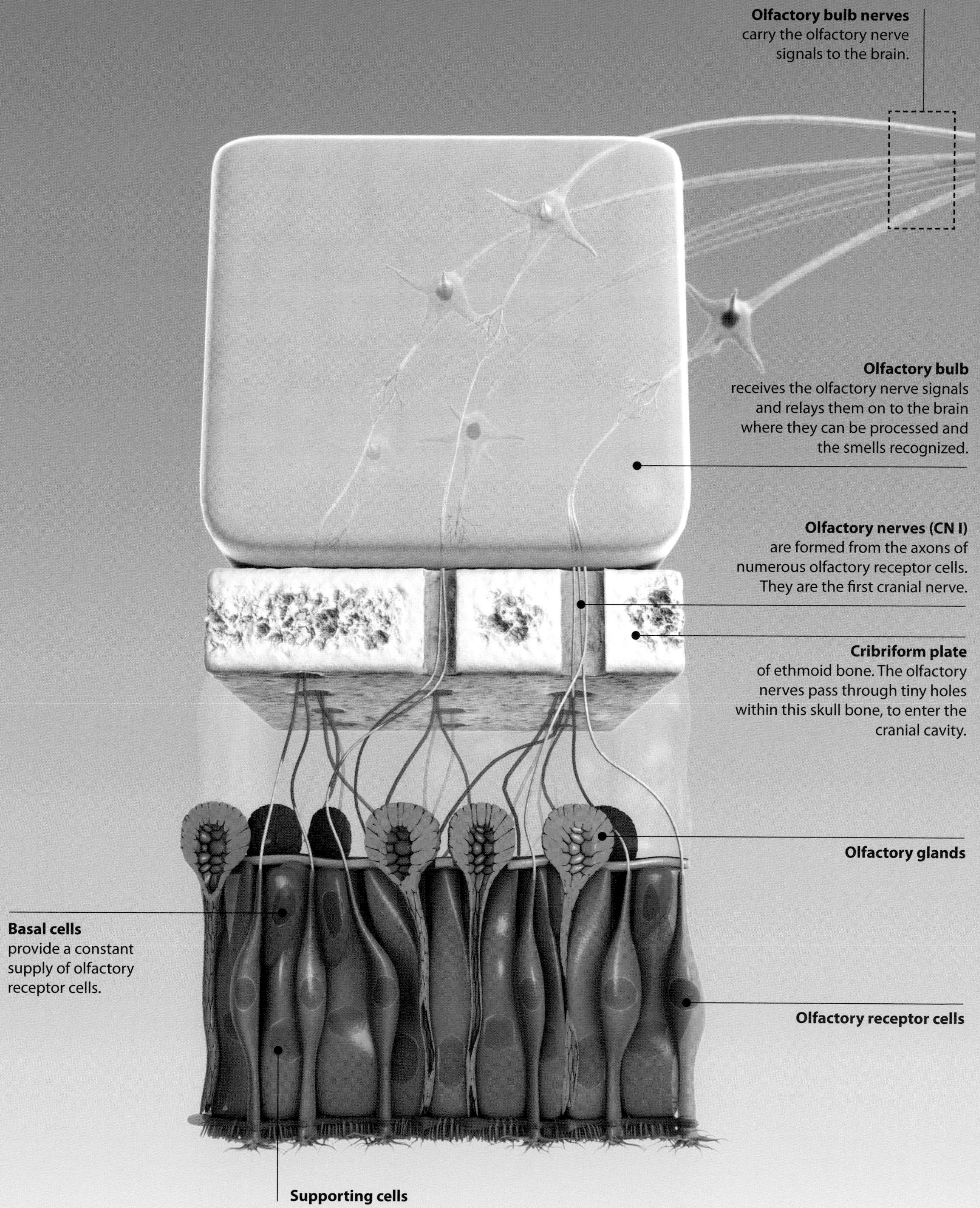

Olfactory bulb nerves
carry the olfactory nerve
signals to the brain.

Olfactory bulb
receives the olfactory nerve signals
and relays them on to the brain
where they can be processed and
the smells recognized.

Olfactory nerves (CN I)
are formed from the axons of
numerous olfactory receptor cells.
They are the first cranial nerve.

Cribriform plate
of ethmoid bone. The olfactory
nerves pass through tiny holes
within this skull bone, to enter the
cranial cavity.

Olfactory glands

Basal cells
provide a constant
supply of olfactory
receptor cells.

Olfactory receptor cells

Supporting cells

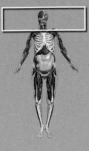

The oral cavity, or mouth, is a passageway through which air and food can reach the throat. The hard and soft palates form the roof, separating it from the nasal cavity, while the tongue forms the floor. The teeth and cheeks form the sides of the oral cavity.

Saliva is a watery secretion that helps keep the mouth moist. It also contains substances that help fight infection, aid digestion, and lubricate food. Saliva helps us taste food by dissolving molecules that can then be detected by the taste buds. The majority of saliva is produced by three paired salivary glands, which drain into the oral cavity.

Did you know?

The salivary glands produce about 1 to 1.5 liters of saliva each day.

Masseter
is a muscle involved in chewing.

Maxilla
forms the upper jaw.

Orbicularis oris

Parotid glands
are found in front of and below the ears. They produce a watery saliva secretion.

Submandibular gland

Buccinator
stops food getting stuck in the cheeks. It is pierced by the parotid duct.

Mandible
is the lower jaw bone.

Parotid duct
carries saliva from the parotid gland to the oral cavity. It pierces the buccinator muscle to enter the mouth.

Parotid Gland

Submandibular and Sublingual Glands

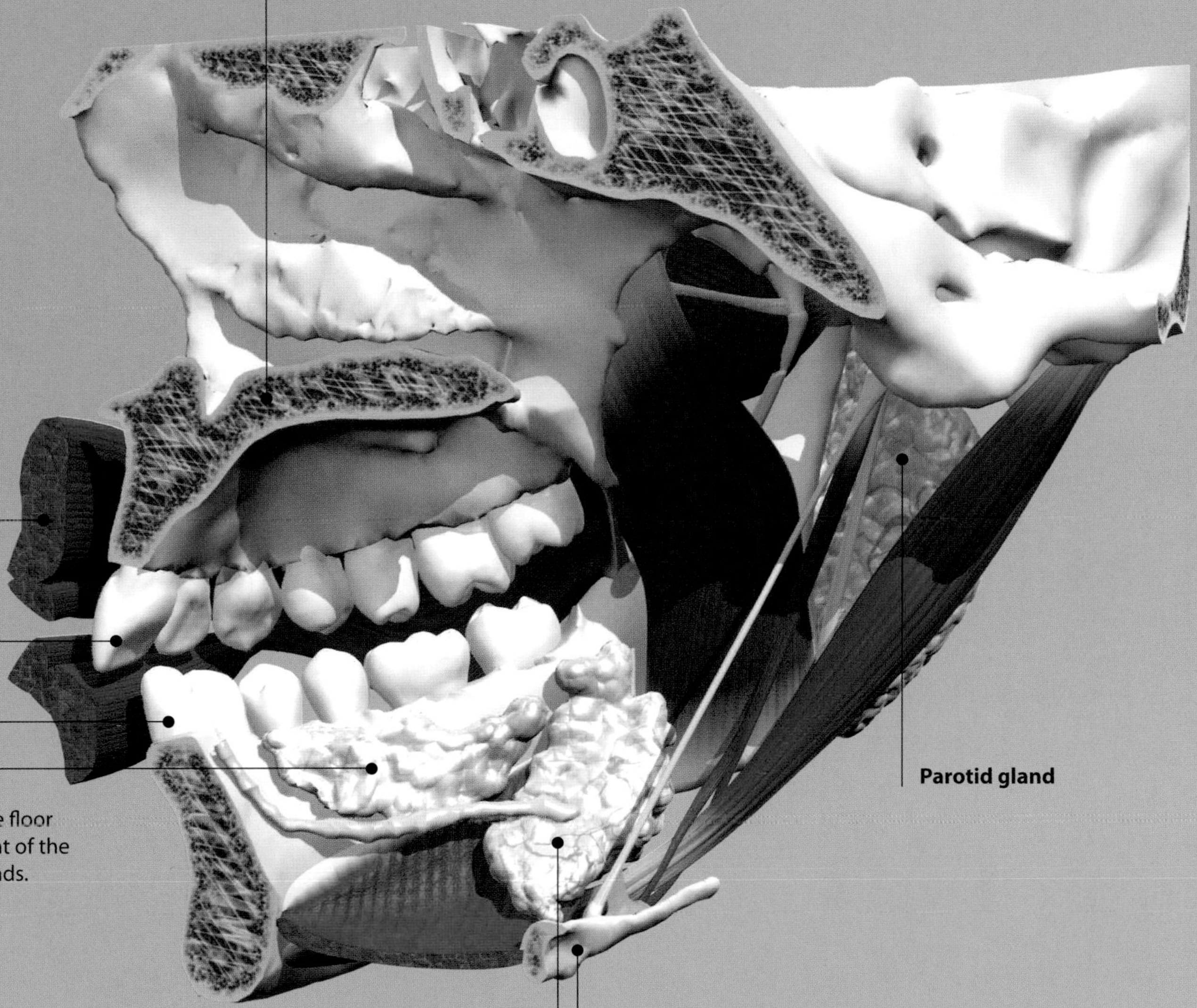

Hard palate
separates the oral
and nasal cavities.

Orbicularis oris muscle
is contained within the
fleshy lips. It surrounds the
entrance to the oral cavity.

Teeth

Sublingual glands
are found within the floor
of the mouth in front of the
submandibular glands.

Parotid gland

Hyoid
is a small bone in the
neck which provides
attachments for
various muscles.

Submandibular glands
lie on the floor of the oral cavity.

Microanatomy of Salivary Gland

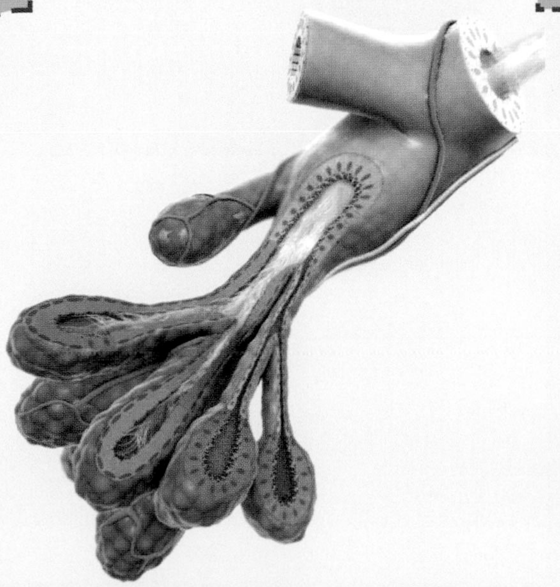

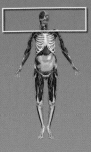

The tongue forms the floor of the oral cavity (mouth), and also extends into part of the pharynx (throat). It is a strong muscular organ, with a roughened top surface and smooth underside. Its top surface has numerous taste buds that allow us to distinguish sweet, salt, sour, bitter, and savory (umami) tastes.

The muscles of the tongue are used to shape and mold food, assist with swallowing, as well as allowing for the delicate, intricate movements required to form words for speech.

Tongue from Above

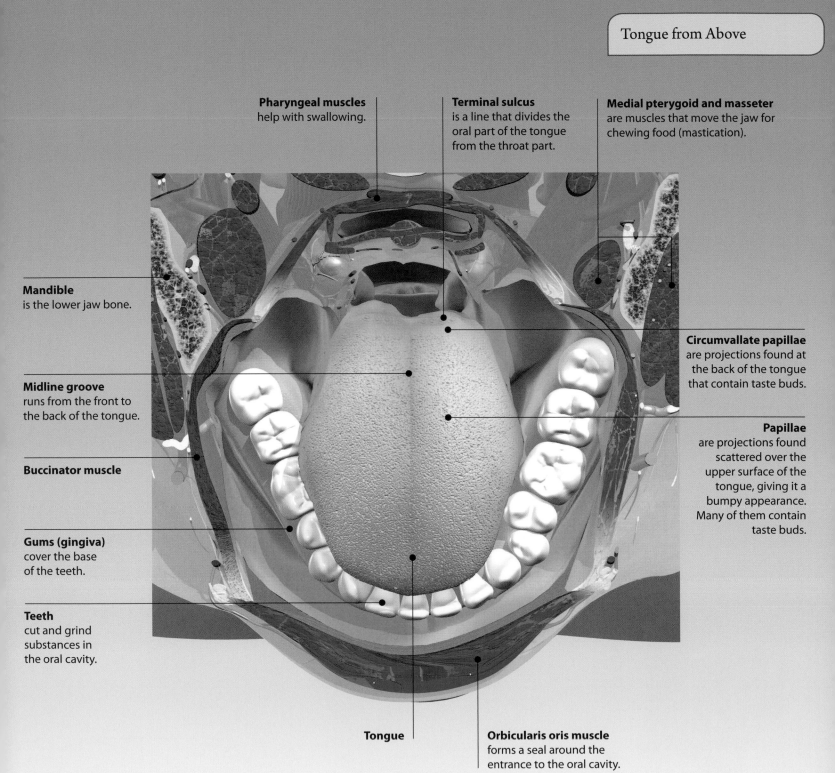

Pharyngeal muscles
help with swallowing.

Terminal sulcus
is a line that divides the oral part of the tongue from the throat part.

Medial pterygoid and masseter
are muscles that move the jaw for chewing food (mastication).

Mandible
is the lower jaw bone.

Midline groove
runs from the front to the back of the tongue.

Buccinator muscle

Gums (gingiva)
cover the base of the teeth.

Teeth
cut and grind substances in the oral cavity.

Circumvallate papillae
are projections found at the back of the tongue that contain taste buds.

Papillae
are projections found scattered over the upper surface of the tongue, giving it a bumpy appearance. Many of them contain taste buds.

Tongue

Orbicularis oris muscle
forms a seal around the entrance to the oral cavity.

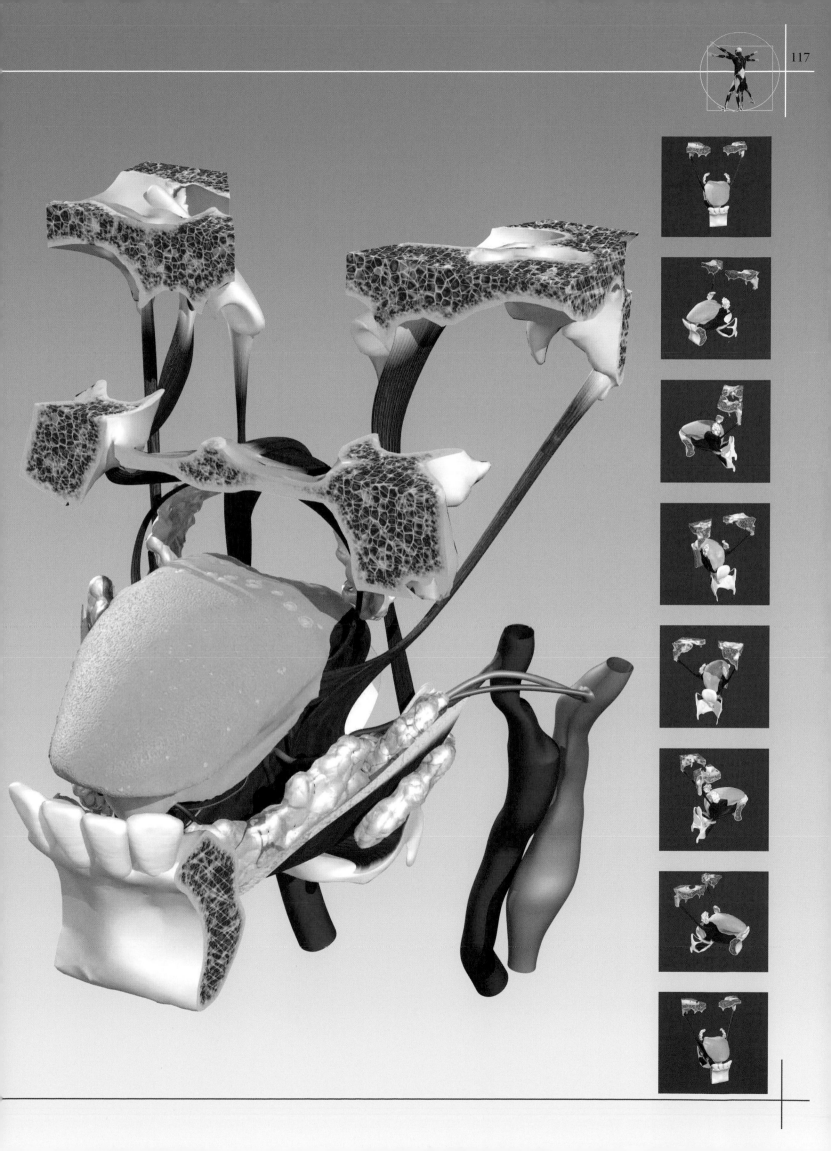

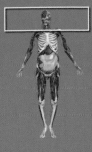

Taste buds are pear-shaped collections of specialized cells, which are able to detect small molecules of food dissolved in the saliva. These molecules trigger a nerve signal to be sent to the brain, which is interpreted as the sensation of taste (gustation). The five main taste types are sweet, salty, savory (umami), bitter, and sour. Taste sensation is carried to the brain in cranial nerves VII (facial) and IX (glossopharyngeal).

We have approximately 10,000 taste buds. Most of these are found on the upper surface of the tongue, related to projections of the tongue surface, called papillae. These give the upper surface of the tongue its roughened appearance.

Filiform papillae
are tall, thin projections, found covering the upper surface of the tongue. They provide a roughened surface to help move food around. They do not have taste buds associated with them.

Foliate papillae
are found along the sides of the tongue.

Circumvallate papillae
form a group at the back of the tongue. They are the largest type of papillae.

Taste buds
are collections of specialized cells that can detect different tastes.

Fungiform papillae
are mushroom-shaped. They are found all over the upper surface of the tongue.

Microanatomy of the Tongue

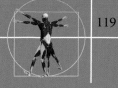

Taste Bud

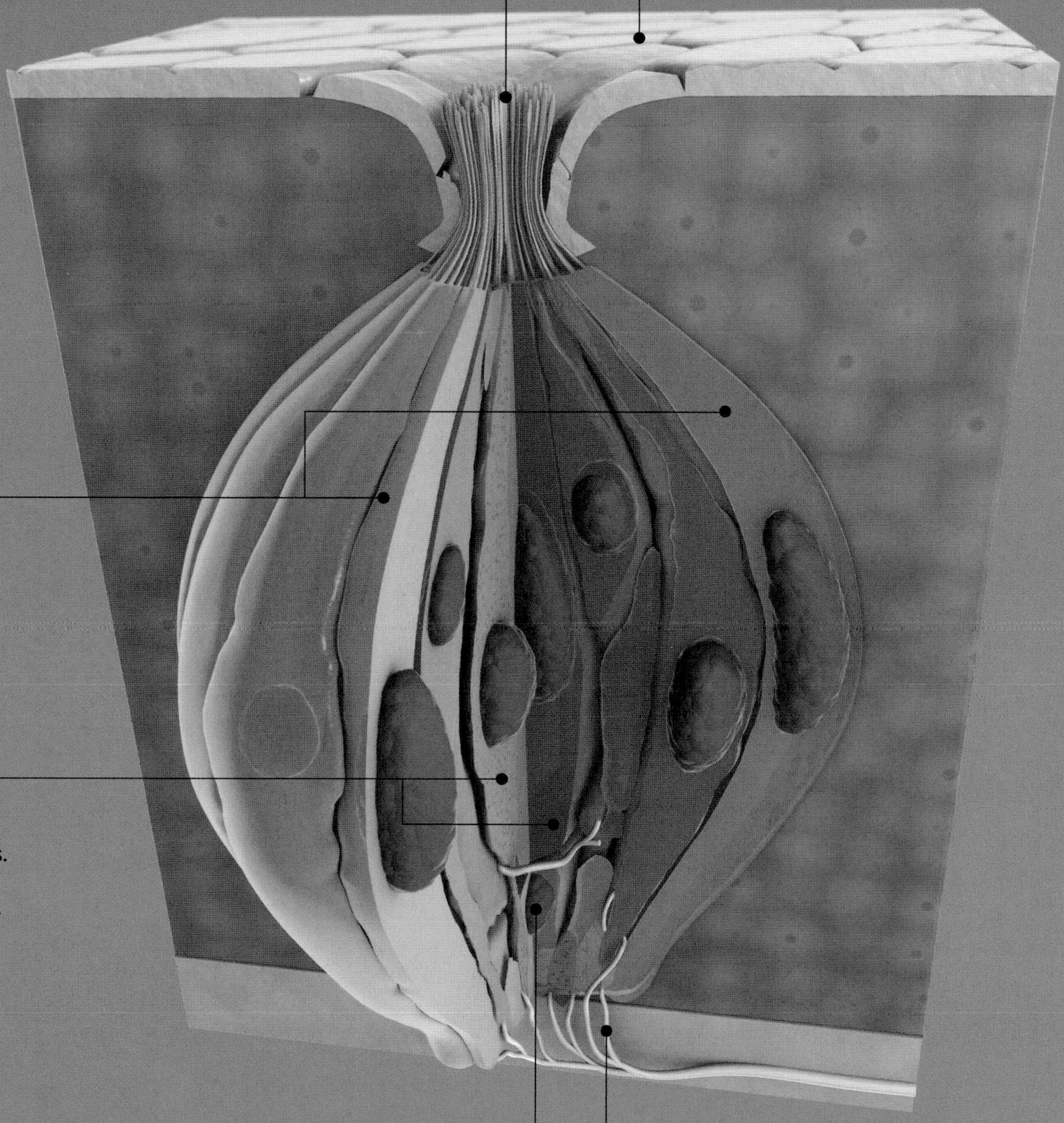

Taste pore
is a small opening onto the
tongue surface. It allows the
contents of the oral cavity to
access the taste buds.

Tongue surface

Supporting cells
surround and support
the gustatory cells.

Gustatory cells
are involved with
sensing different tastes.
They have a lifespan of
approximately 10 days.

Basal cells
are small cells found at the
bottom of the taste bud. They
provide a constant supply of
gustatory and supporting cells.

Nerve fibers
carry the taste signals to the brain.

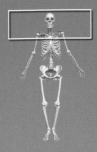

Within the oral cavity are two rows of teeth. They are embedded within sockets in the upper and lower jaw bones. Their role is to break food down into smaller pieces that can be more easily digested. Based on their shape and site, there are four main types of teeth: incisors, canines, premolars, and molars. Deciduous or "milk" teeth first appear from age 6 months. There are 20 deciduous teeth in total. From about six years of age until early adulthood, the milk teeth are gradually replaced by permanent teeth. In total, there are 32 permanent teeth: 16 in both the upper and lower jaw.

Pharynx
delivers food to the digestive tract where further processing takes place.

Tongue

Submandibular salivary gland

Molars
grind food between their cusps. There are 3 molars on each side. The molars found right at the back are known as wisdom teeth, and do not always fully emerge.

Premolars
have projections called cusps that are designed to tear and grind food. There are two premolars on each side, just in front of the molars.

Incisors

Canines

Mandible (lower jaw bone)

Mandibular Teeth

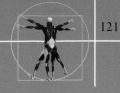

Maxillary Teeth

Hard palate

Incisors
are the chisel-shaped front teeth. Their sharp edge cuts through food. There are two incisors on each side.

Canines
are pointed teeth which pierce and hold onto food. One canine is found on each side behind the incisors.

Premolars

Molars

Greater palatine nerves and vessels
innervate and supply blood to the gums and hard palate.

Palatine glands
keep the hard palate moist and well lubricated.

Palatine tonsils
are collections of lymphoid material at the back of the mouth that help fight infection.

Uvula
is the small fleshy part of the soft palate that hangs down at the back of the mouth.

Neurovascular Supply

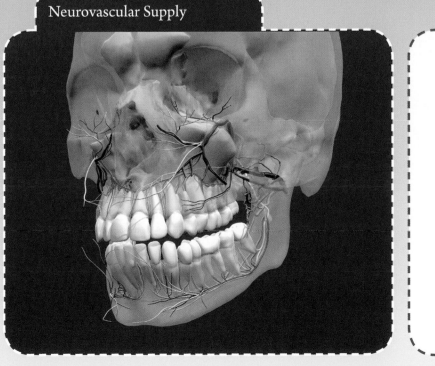

FDI World Dental Federation Notation

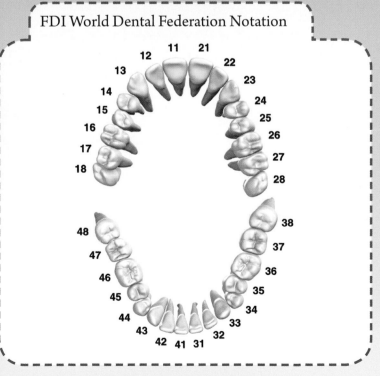

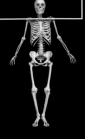

Although there are different shapes and types of teeth, they all have the same basic structure and are made of the same materials. They all have a crown, neck, root, and pulp cavity and are all made from enamel, dentin, cementum, and pulp.

The structure of teeth can be damaged by exposure to bacteria and sugars. These can break down the tough outer enamel layer leading to cavities, and possibly painful dental abscesses. Prompt treatment is needed to drain the infection and seal the holes with a filling.

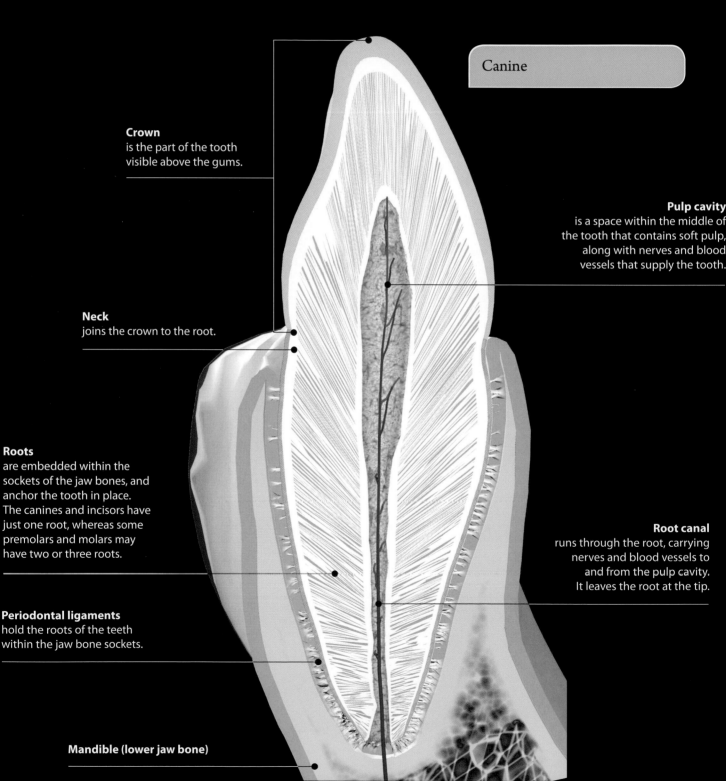

Canine

Crown
is the part of the tooth visible above the gums.

Pulp cavity
is a space within the middle of the tooth that contains soft pulp, along with nerves and blood vessels that supply the tooth.

Neck
joins the crown to the root.

Roots
are embedded within the sockets of the jaw bones, and anchor the tooth in place. The canines and incisors have just one root, whereas some premolars and molars may have two or three roots.

Root canal
runs through the root, carrying nerves and blood vessels to and from the pulp cavity. It leaves the root at the tip.

Periodontal ligaments
hold the roots of the teeth within the jaw bone sockets.

Mandible (lower jaw bone)

Enamel
forms the crown. It is the hardest material found in the human body.

Crown

Dentin
is similar to bone and forms most of the basic structure of the tooth.

Gingiva (gums)
form a fleshy support around the neck and root of each tooth.

Neck

Pulp cavity

Pulp
is a soft material found within the pulp cavity. It surrounds the blood vessels and nerves running in the pulp cavity and root canal.

Cementum
helps anchor the roots to the bony jaw bone sockets.

Roots

Periodontal ligaments

Root canal

Mandible

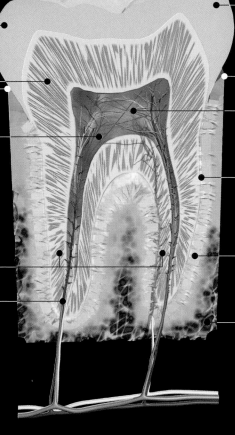

Molar

Disease

Abscess

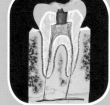

Pulp penetrating decay

Cavities

Pulpitis

Gingivitis and peridontal disease

Restoration

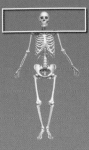

The oral cavity and teeth are innervated by a number of different nerves. The transmission of pain sensation down these nerves can be blocked by injecting local anesthetic agents around them. This allows dental procedures to be carried out in a pain-free manner.

The choice of which intraoral injection to use depends mainly upon the area that needs to be pain free (anesthetized). All of these injections require excellent knowledge of the relevant anatomy to allow successful anesthesia.

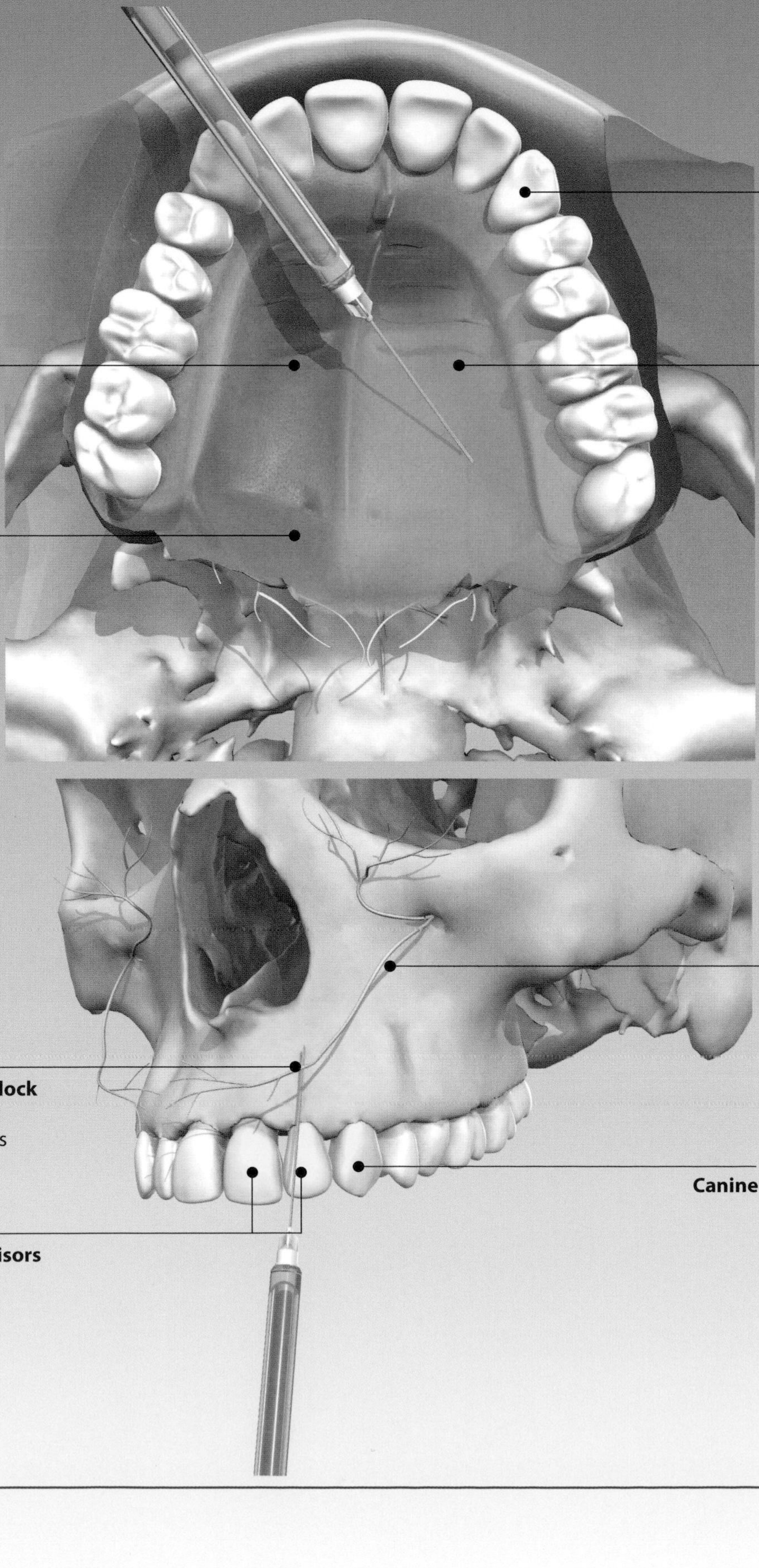

Canine

Hard palate

Greater palatine nerve block leads to anesthesia of the hard and soft palate behind the canine on the injected side.

Soft palate

Maxillary nerve

Anterior superior alveolar nerve block anesthetizes around the incisors and canine on the injected side, as well as the overlying lip.

Canine

Incisors

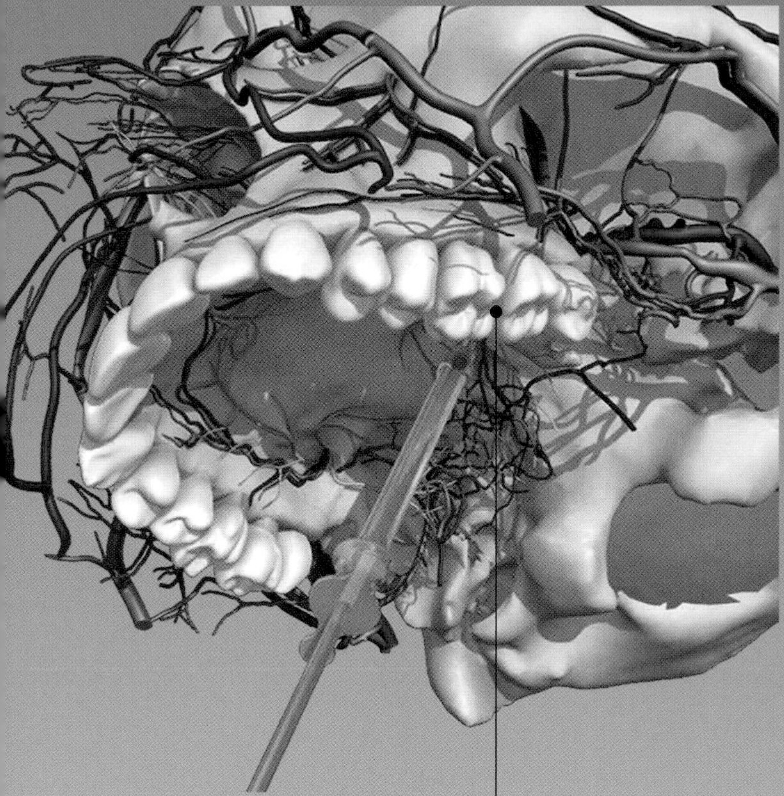

Maxillary division nerve block
anesthetizes the maxillary
nerve before it can give off any
branches. This provides anesthesia
to all the upper teeth and palate
on the injected side.

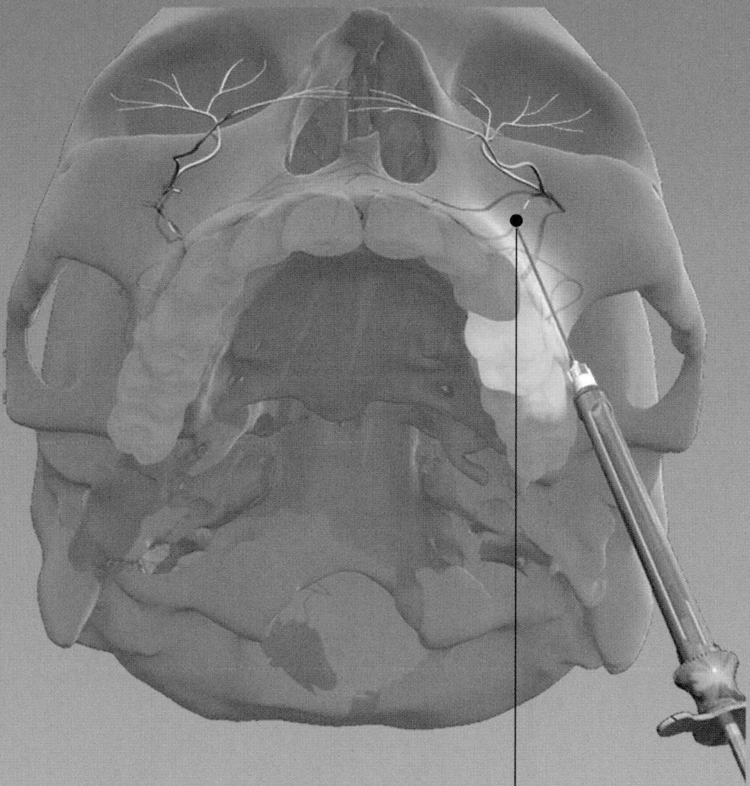

Middle superior alveolar nerve block
anesthetizes the premolars and overlying gum.

Buccinator

Inferior alveolar nerve

Inferior alveolar nerve block
anaesthetizes all of the teeth in
the mandible on the injected side.
In addition, the skin over the chin
and lower lips on the injected side
are also affected.

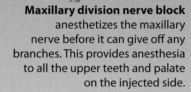

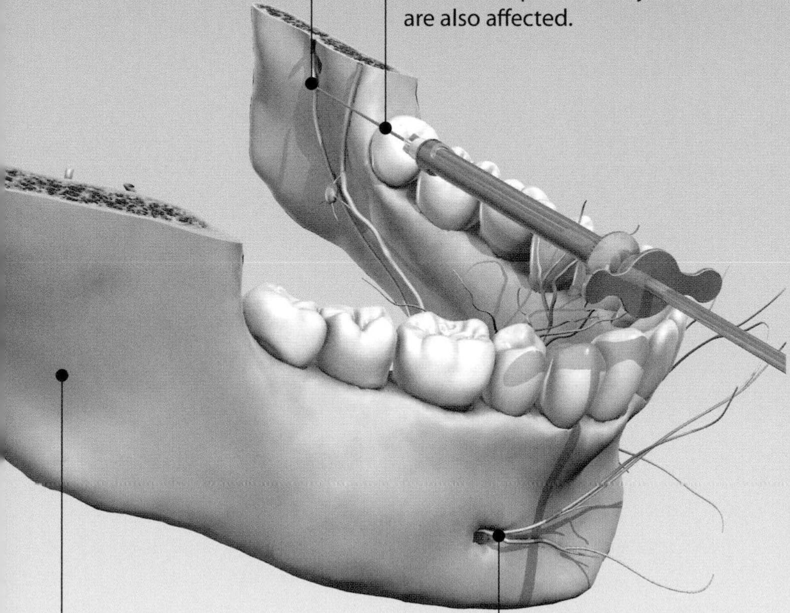

Mandible

Mental nerve
is a continuation of the inferior
alveolar nerve. It supplies the
skin over the chin.

Long buccal nerve block
anesthetizes the gums
around the molars along
with the skin over the cheek.

Masseter

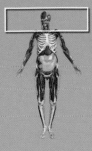

The pharynx, or throat, is a muscular passageway for food, liquids, and air. It connects the oral and nasal cavities to the esophagus and larynx. The pharynx is divided into three sections: the nasopharynx, oropharynx, and laryngopharynx.

Three pharyngeal constrictor muscles form most of the pharyngeal walls. They partially overlap each other to form a muscular tube, which helps move food and fluid from the pharynx into the esophagus.

Nasopharynx
is located behind the nasal cavity. It carries air to the oropharynx, which it joins at the level of the soft palate. It contains the openings of the auditory tubes, which connect it to the middle ear. They allow the air pressure on either side of the eardrum to equalize.

Nasal cavity

Oral cavity

Tongue

Soft palate

Palatine tonsils
are collections of lymphoid tissue in the oropharynx that help fight infection.

Oropharynx
is found at the back of the oral cavity. It carries both food and air, and joins the laryngopharynx below at the level of the hyoid bone.

Epiglottis
is a mobile piece of cartilage that stops food going into the larynx.

Hyoid bone
is a small bone in the neck. Numerous muscles involved with moving the tongue and swallowing are attached to it.

Larynx (voice box)

Laryngopharynx
opens into the larynx and esophagus at its lower end.

Pharyngeal constrictor muscles
help move food and fluids from the pharynx into the esophagus.

Pharynx and Larynx

Trachea,
or windpipe, is found in front of the esophagus.

Esophagus
(gullet) is a muscular tube that carries food to the stomach.

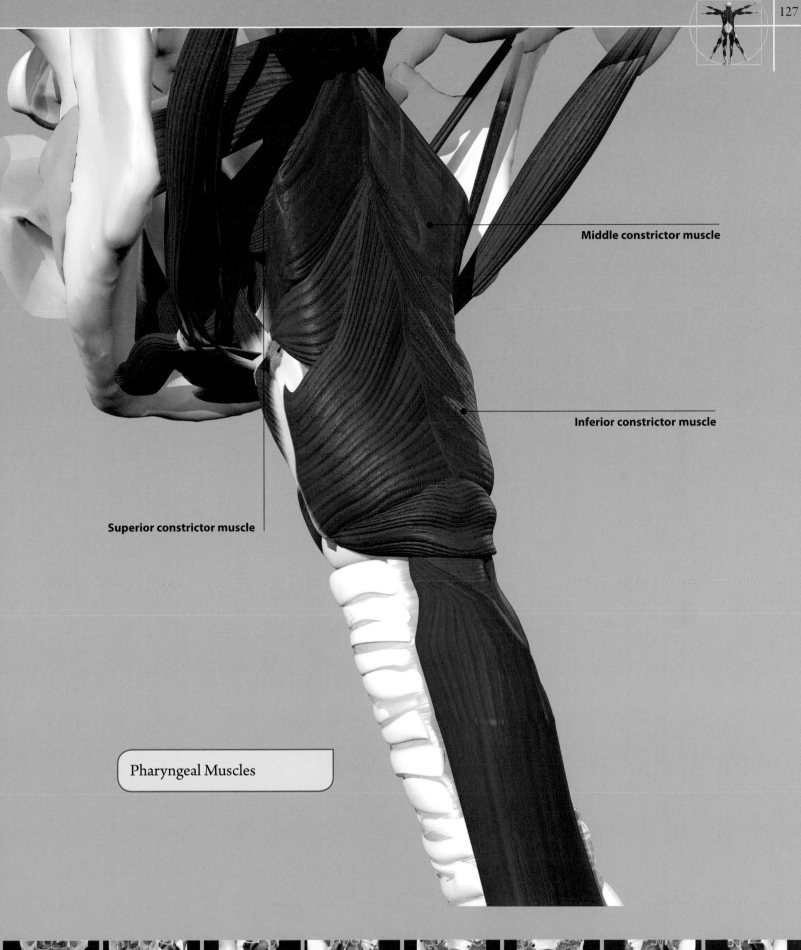

Middle constrictor muscle

Inferior constrictor muscle

Superior constrictor muscle

Pharyngeal Muscles

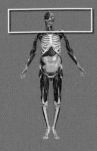

The larynx, or voice box, is found at the front of the neck. It is a short passageway through which air can move from the laryngopharynx into the trachea (windpipe). It is made up of a number of separate pieces of cartilage, held together by connective tissue and muscles.

Paired pieces of tissue called "vocal folds" (or vocal cords) are attached between the cartilages of the larynx. Air passing between them causes them to vibrate and produce sounds that help us speak. Muscles that move the cartilage of the larynx alter the position and length of the vocal cords, to form sounds of different pitch.

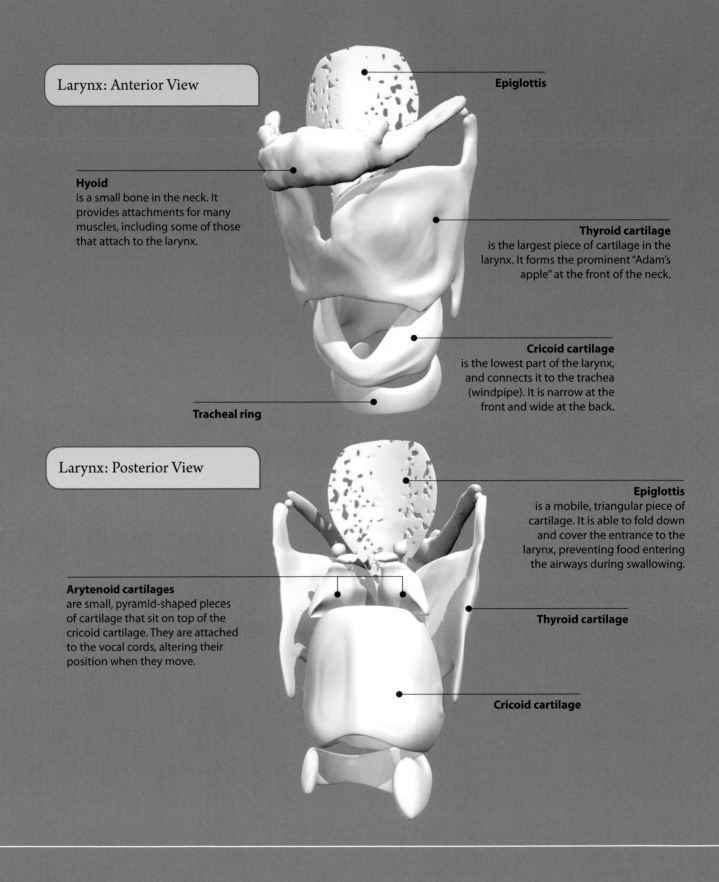

Larynx: Anterior View

Epiglottis

Hyoid
is a small bone in the neck. It provides attachments for many muscles, including some of those that attach to the larynx.

Thyroid cartilage
is the largest piece of cartilage in the larynx. It forms the prominent "Adam's apple" at the front of the neck.

Cricoid cartilage
is the lowest part of the larynx, and connects it to the trachea (windpipe). It is narrow at the front and wide at the back.

Tracheal ring

Larynx: Posterior View

Epiglottis
is a mobile, triangular piece of cartilage. It is able to fold down and cover the entrance to the larynx, preventing food entering the airways during swallowing.

Arytenoid cartilages
are small, pyramid-shaped pieces of cartilage that sit on top of the cricoid cartilage. They are attached to the vocal cords, altering their position when they move.

Thyroid cartilage

Cricoid cartilage

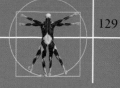

Larynx: Superior View

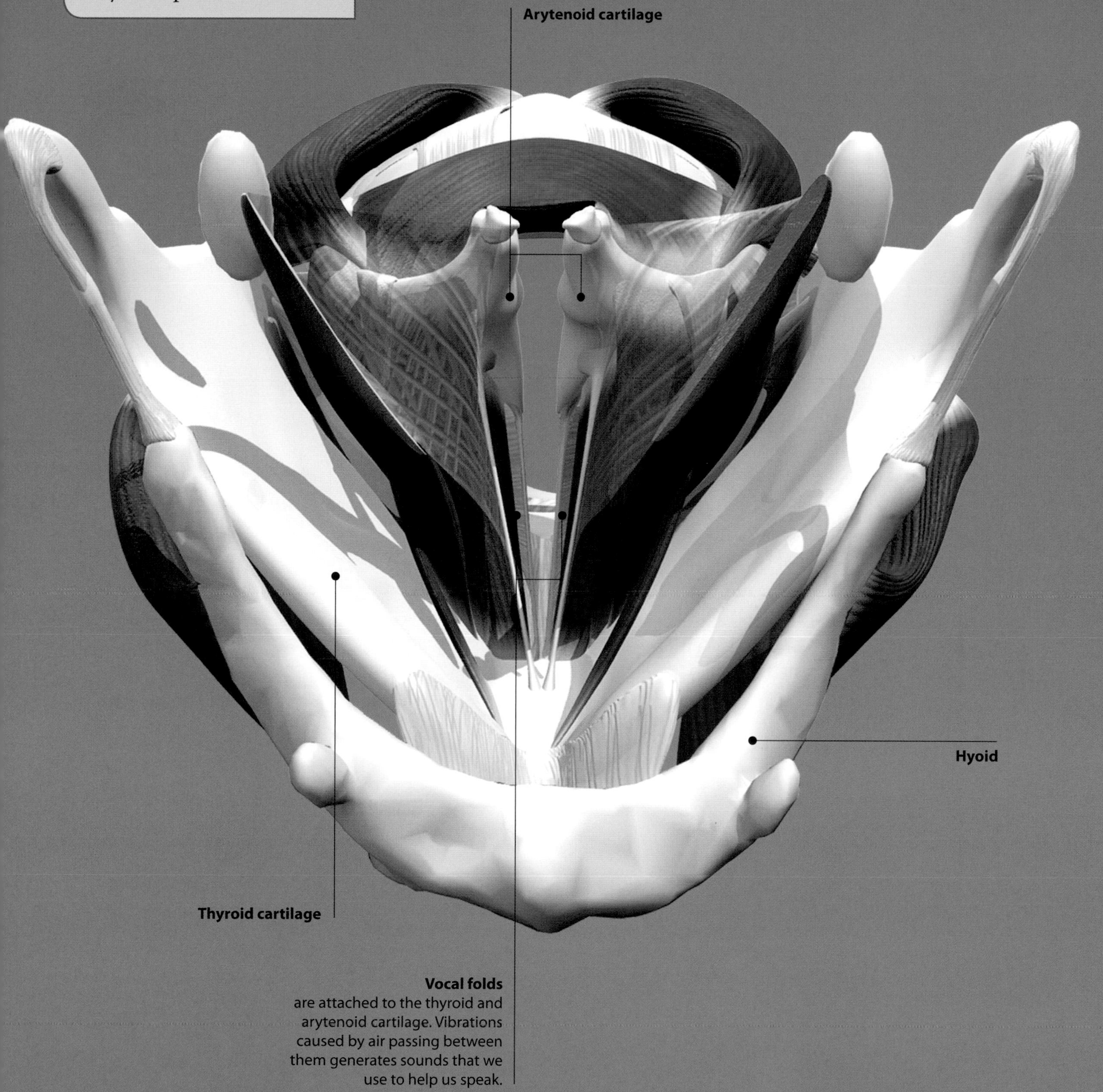

Arytenoid cartilage

Hyoid

Thyroid cartilage

Vocal folds
are attached to the thyroid and
arytenoid cartilage. Vibrations
caused by air passing between
them generates sounds that we
use to help us speak.

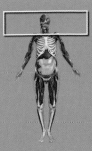

Swallowing is the process by which food and fluids are moved from the mouth to the stomach. It involves the coordinated contraction of muscles of the tongue, soft palate, pharynx, larynx, and esophagus. The process is controlled by the nervous system. There are three phases to swallowing: oral, pharyngeal, and esophageal.

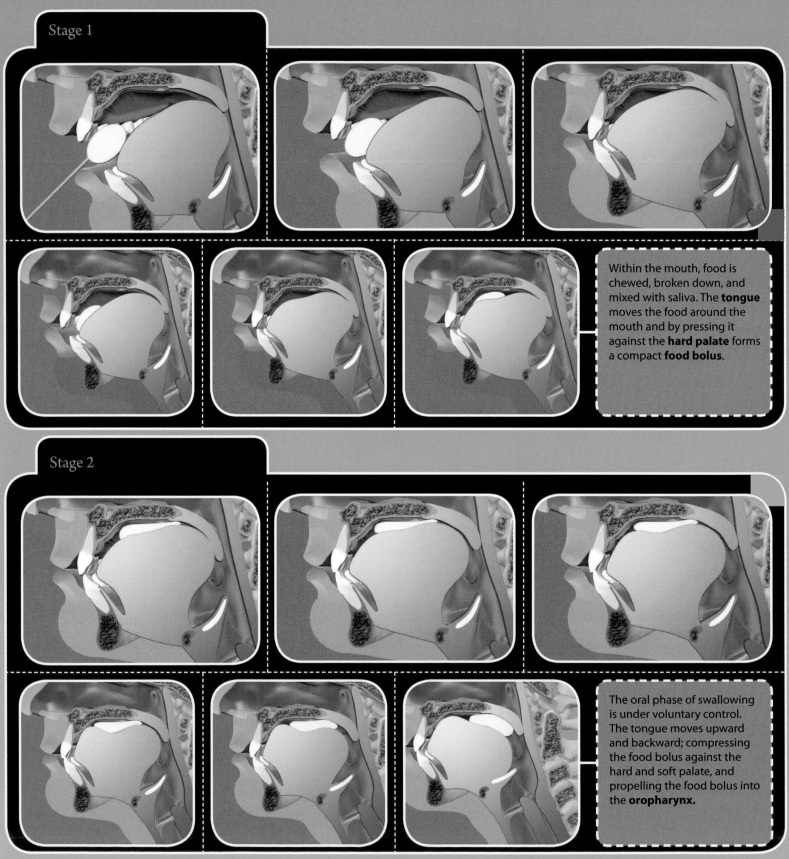

Stage 1

Within the mouth, food is chewed, broken down, and mixed with saliva. The **tongue** moves the food around the mouth and by pressing it against the **hard palate** forms a compact **food bolus**.

Stage 2

The oral phase of swallowing is under voluntary control. The tongue moves upward and backward; compressing the food bolus against the hard and soft palate, and propelling the food bolus into the **oropharynx.**

Stage 3

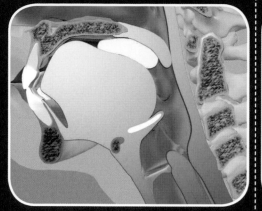

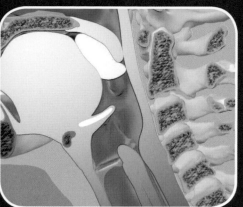

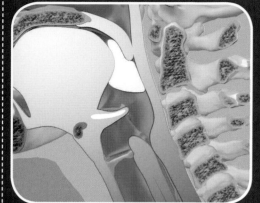

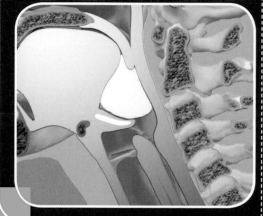

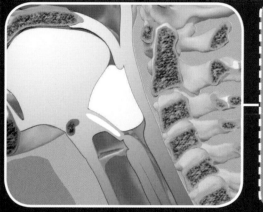

The pharyngeal phase is not under conscious voluntary control. When the food bolus enters the oropharynx, the swallowing center in the brainstem triggers a reflex contraction of muscles of the soft palate, pharynx, and larynx. They move the bolus through the pharynx to the esophagus opening. The soft palate moves upward to prevent food entering the nasopharynx. The epiglottis folds down and larynx moves up, preventing food from entering the trachea.

Stage 4

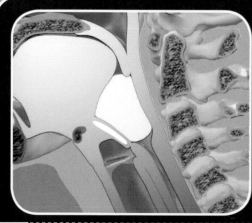

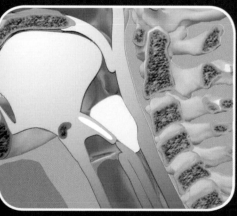

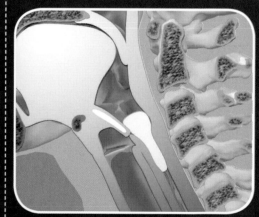

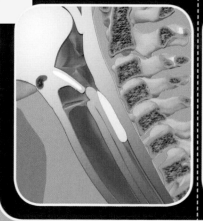

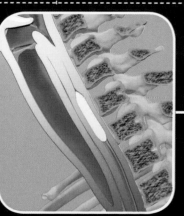

The esophageal phase is also involuntary. A series of coordinated wave-like muscle contractions (**peristaltic waves**) sweep the food bolus down the **esophagus** and into the stomach. This process is known as peristalsis.

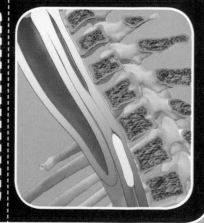

BONES OF THE NECK

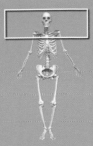

The skull is supported by seven bones called cervical vertebrae (referred to as C1 to C7). They form the highest section of the vertebral column. The vertebrae are stacked on top of each other, forming a canal through which the spinal cord can run.

They allow the head to move in a variety of directions: bending forward and backward and side to side, and rotating the head both left and right.

Skull

C1 (atlas)
connects the base of the skull to the vertebral column.

C2 (axis)
is the second cervical vertebra and lies beneath the atlas.

C4
is the fourth cervical vertebra.

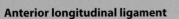

Anterior longitudinal ligament

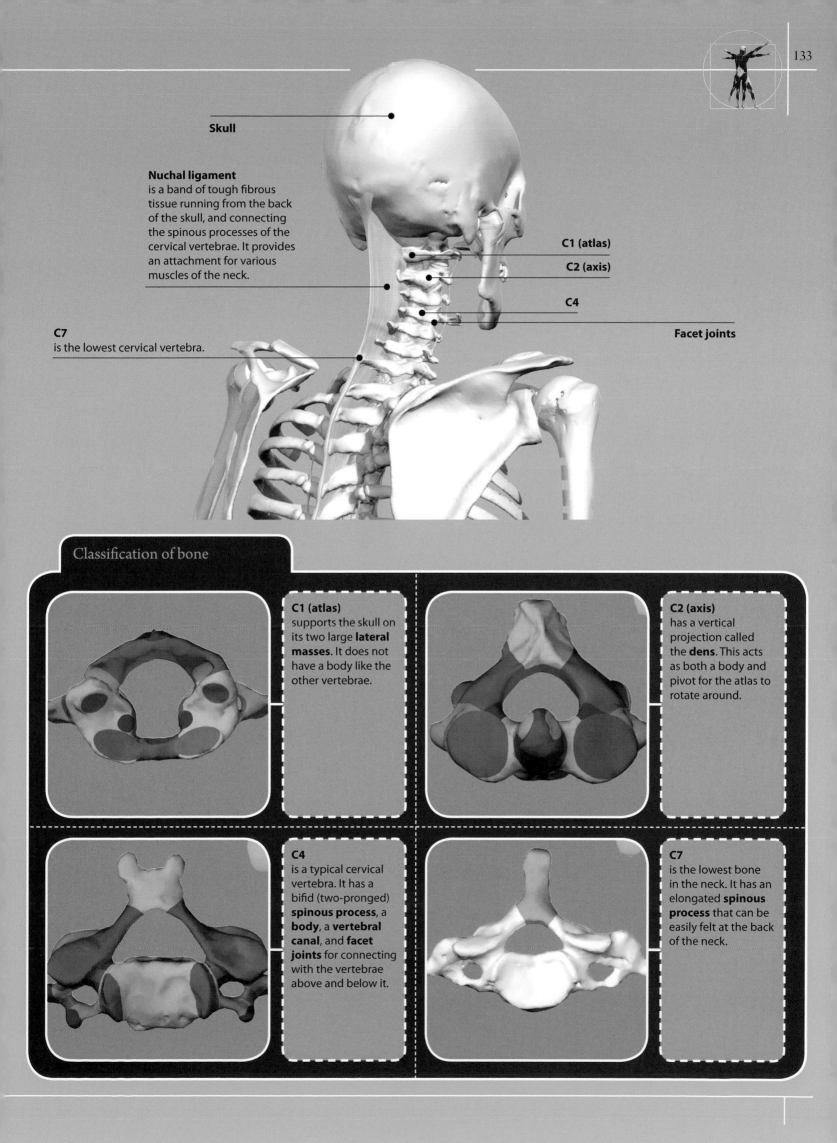

Skull

Nuchal ligament
is a band of tough fibrous tissue running from the back of the skull, and connecting the spinous processes of the cervical vertebrae. It provides an attachment for various muscles of the neck.

C1 (atlas)

C2 (axis)

C4

C7
is the lowest cervical vertebra.

Facet joints

Classification of bone

C1 (atlas)
supports the skull on its two large **lateral masses**. It does not have a body like the other vertebrae.

C2 (axis)
has a vertical projection called the **dens**. This acts as both a body and pivot for the atlas to rotate around.

C4
is a typical cervical vertebra. It has a bifid (two-pronged) **spinous process**, a **body**, a **vertebral canal**, and **facet joints** for connecting with the vertebrae above and below it.

C7
is the lowest bone in the neck. It has an elongated **spinous process** that can be easily felt at the back of the neck.

Movements of the neck and the position of the head are controlled by a large number of muscles. Stability and fine adjustment tends to be performed by the small muscles found deep within the neck. The large muscles closer to the skin surface are involved with bending the neck forward, backward, and side to side, as well as turning it to the left and right.

Nuchal ligament
is a band of tough fibrous tissue that provides attachment for many of the muscles of the neck.

Rectus capitis posterior muscles
help keep the head upright and facing forward.

Spinalis capitis

Semispinalis capitis

Obliquus capitis muscles
are small muscles that help stabilize the position of the head.

Scalenes
are a group of muscles that bend the neck forward, and rotate it from side to side.

Longissmus cervicis
is a small muscle that bends the neck backward and from side to side.

Iliocostalis cervicis
helps bend the neck backward and to the sides.

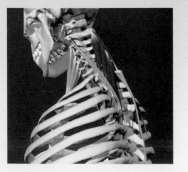
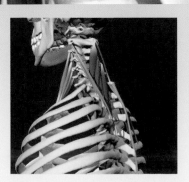
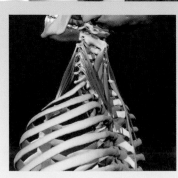
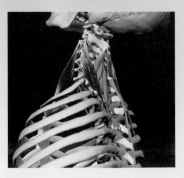

Spinalis capitis
bends the neck backward
and from side to side.

Semispinalis capitis
bends the neck backward
and from side to side.

Levator scapulae
is attached to the base of
the skull. It pulls the scapula
(shoulder blade) upward.

Splenius capitis
is a large muscle at the back
of the neck. It tilts the head
backward when it contracts.

Scapula (shoulder blade)

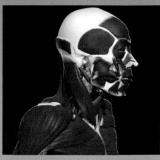

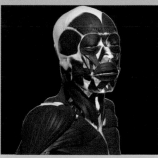

Mastoid process
is part of the skull, just
below and behind the ears.

Sternocleidomastoid muscles
attach to the sternum
(breastbone), clavicle (collar
bone), and mastoid process on
each side. They bend the neck
forward and turn it to the sides.

Trapezius
is a large diamond-
shaped muscle that
is attached to the
neck and the back. It
raises the shoulders,
as in shrugging.

Individual nerves from different levels of the spinal cord join together to form a network of fibers. This is called a plexus. Branches from the plexus then supply the skin and muscles of a specific area.

The cervical plexus is formed from the upper four spinal nerves in the neck (cervical nerves C1-C4). The plexus gives off nerve branches that supply the skin of the shoulders, neck and scalp, the muscles at the front of the neck, as well as the diaphragm, the main muscle that helps us breathe.

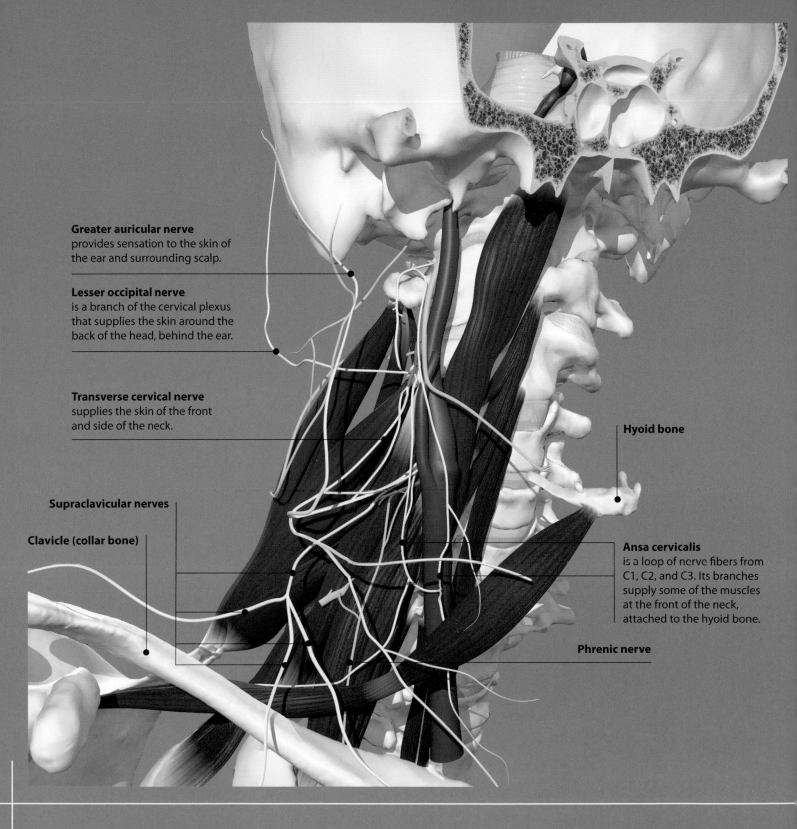

Greater auricular nerve
provides sensation to the skin of the ear and surrounding scalp.

Lesser occipital nerve
is a branch of the cervical plexus that supplies the skin around the back of the head, behind the ear.

Transverse cervical nerve
supplies the skin of the front and side of the neck.

Supraclavicular nerves

Clavicle (collar bone)

Hyoid bone

Ansa cervicalis
is a loop of nerve fibers from C1, C2, and C3. Its branches supply some of the muscles at the front of the neck, attached to the hyoid bone.

Phrenic nerve

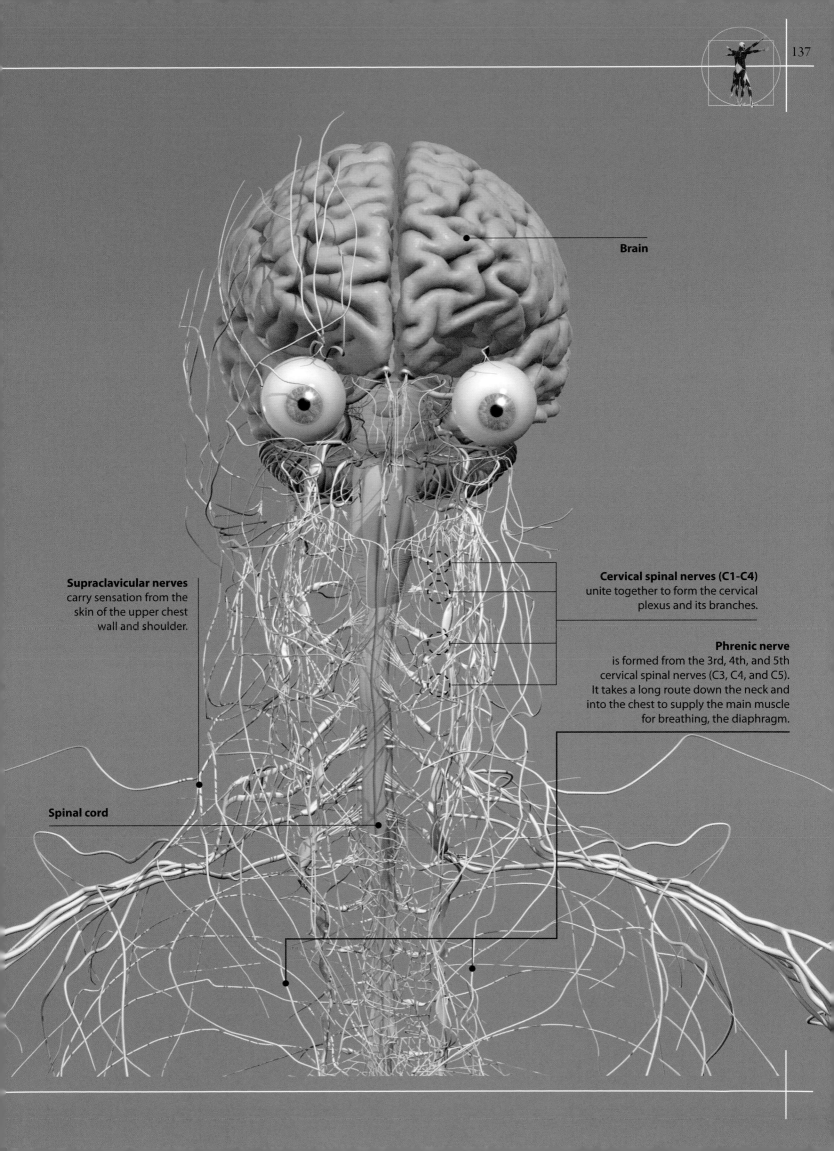

Brain

Supraclavicular nerves carry sensation from the skin of the upper chest wall and shoulder.

Cervical spinal nerves (C1-C4) unite together to form the cervical plexus and its branches.

Phrenic nerve is formed from the 3rd, 4th, and 5th cervical spinal nerves (C3, C4, and C5). It takes a long route down the neck and into the chest to supply the main muscle for breathing, the diaphragm.

Spinal cord

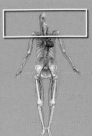

The neck is tightly packed with muscles, blood vessels, nerves, and other important structures.

Did you know?

The internal jugular vein is fairly consistent in position between individuals. Large tubes called central venous catheters can be introduced into this vein by healthcare professionals, to help them monitor and give fluids to critically ill patients.

Sublingual salivary glands

External carotid artery
supplies the face and neck.

Carotid sinus
is a widened area found where the common carotid splits into its external and internal branches. It helps in the control of blood pressure.

Internal jugular vein
drains most of the blood from the head and neck.

Thyroid gland
is firmly attached to the front of the trachea. It is an endocrine organ that produces thyroid hormones.

Scalene muscles

Hyoid bone

Internal carotid artery
enters the skull to supply most of the brain.

Thyroid cartilage
is the prominent part of the larynx (voice box) known as the "Adam's apple."

Common carotid artery
is the main source of blood to the head and neck. Its pulsations can be felt next to the thyroid cartilage.

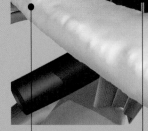

Clavicle
(collar bone)

Phrenic nerve
travels down into the chest to supply the diaphragm.

Trachea
(windpipe)

Cervical Plexus

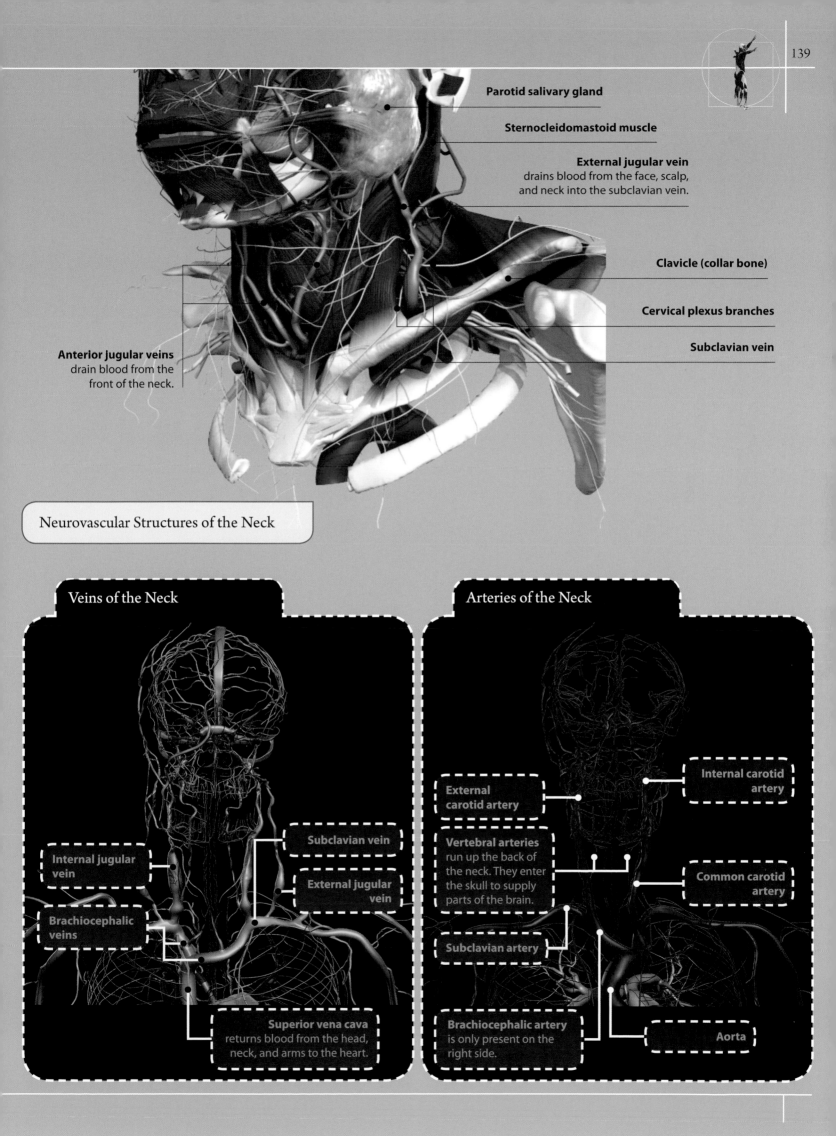

Parotid salivary gland

Sternocleidomastoid muscle

External jugular vein
drains blood from the face, scalp, and neck into the subclavian vein.

Clavicle (collar bone)

Cervical plexus branches

Subclavian vein

Anterior jugular veins
drain blood from the front of the neck.

Neurovascular Structures of the Neck

Veins of the Neck

Subclavian vein

External jugular vein

Internal jugular vein

Brachiocephalic veins

Superior vena cava
returns blood from the head, neck, and arms to the heart.

Arteries of the Neck

External carotid artery

Internal carotid artery

Vertebral arteries
run up the back of the neck. They enter the skull to supply parts of the brain.

Common carotid artery

Subclavian artery

Brachiocephalic artery
is only present on the right side.

Aorta

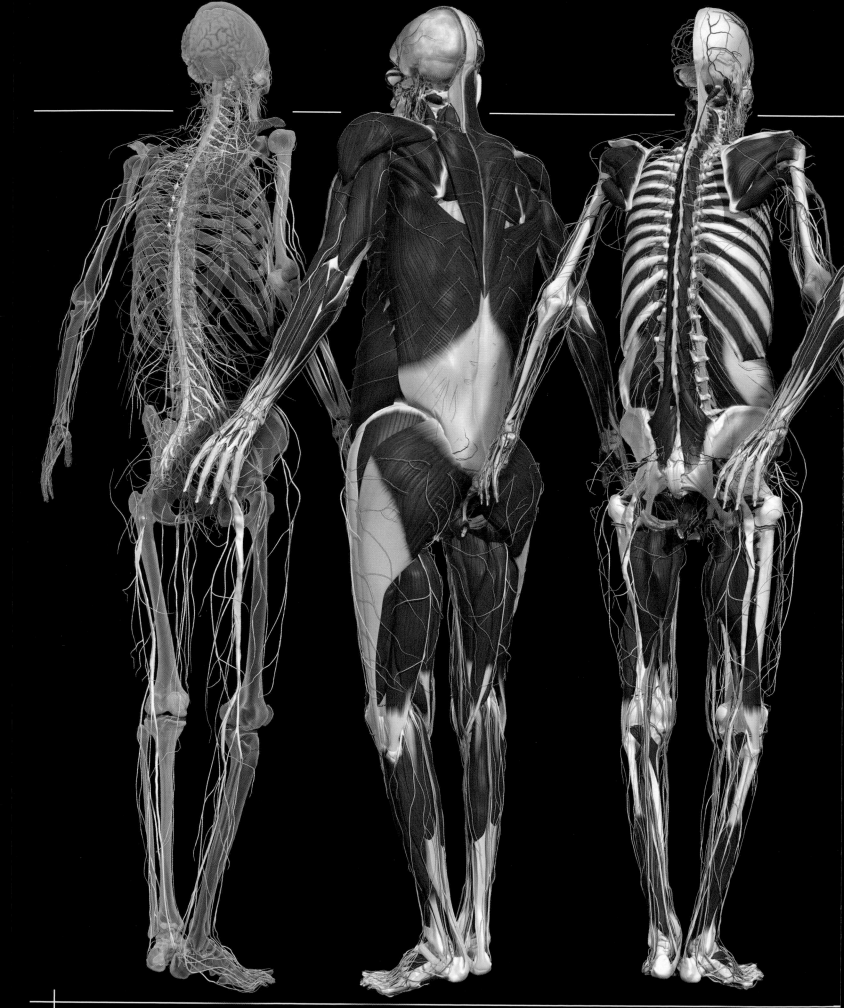

The central part of the body between the neck and the hips is known as the trunk.

The rear most part of this area, which provides the trunk with structure and support, is called the back.

The vertebral column (spine or backbone) forms most of the bony framework of the back, with contributions from the ribs, shoulder blades, and pelvis. It is made up of a number of small bones called vertebrae, which are stacked together to form a chain. This arrangement gives strength to the spine, while still allowing it to move.

Numerous muscles are attached to the vertebral column. They help us keep a good posture, and allow us to bend and twist the spine.

Running through the center of the vertebral column, and protected by it, is the spinal cord. This vital structure relays nerve signals between the brain and rest of the body. Spinal nerves leave the spinal cord at different levels to supply the body.

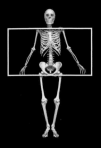

The vertebral column (backbone or spine) forms a strong central axis for the trunk. It is made up of thirty-three small bones called vertebrae, which are stacked together with intervertebral discs between them, to form a strong but mobile column.

The vertebral column is divided into five different regions: cervical, thoracic, lumbar, sacral, and coccygeal. The vertebrae from each region have slightly different features. The normal vertebral column gently curves forward and backward (in the sagittal plane) along its length.

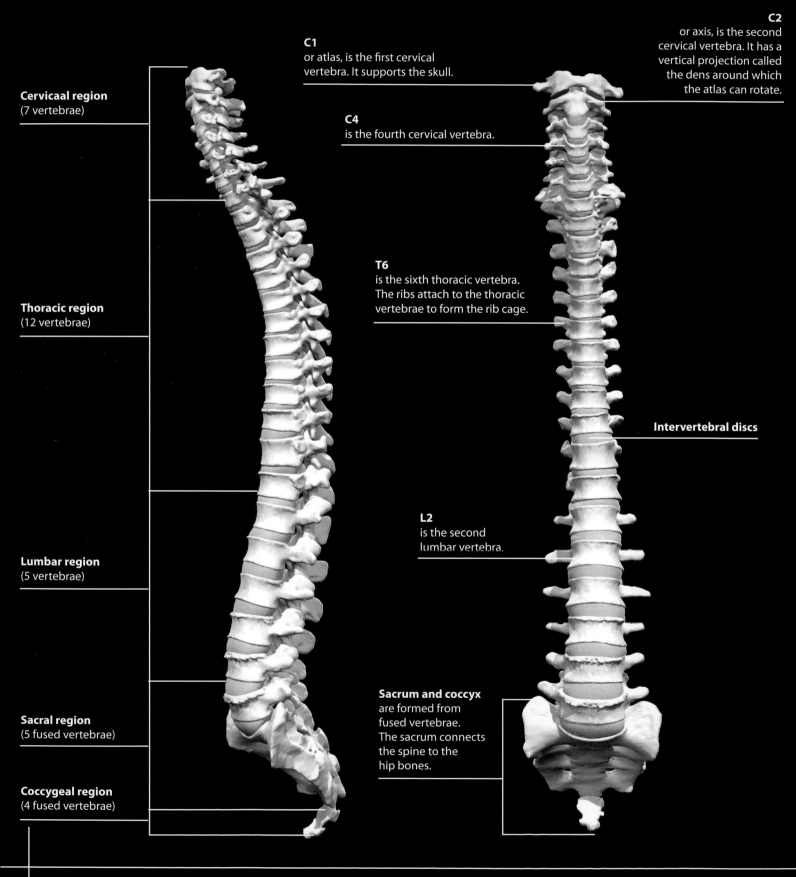

Cervicaal region
(7 vertebrae)

Thoracic region
(12 vertebrae)

Lumbar region
(5 vertebrae)

Sacral region
(5 fused vertebrae)

Coccygeal region
(4 fused vertebrae)

C1
or atlas, is the first cervical vertebra. It supports the skull.

C2
or axis, is the second cervical vertebra. It has a vertical projection called the dens around which the atlas can rotate.

C4
is the fourth cervical vertebra.

T6
is the sixth thoracic vertebra. The ribs attach to the thoracic vertebrae to form the rib cage.

Intervertebral discs

L2
is the second lumbar vertebra.

Sacrum and coccyx
are formed from fused vertebrae. The sacrum connects the spine to the hip bones.

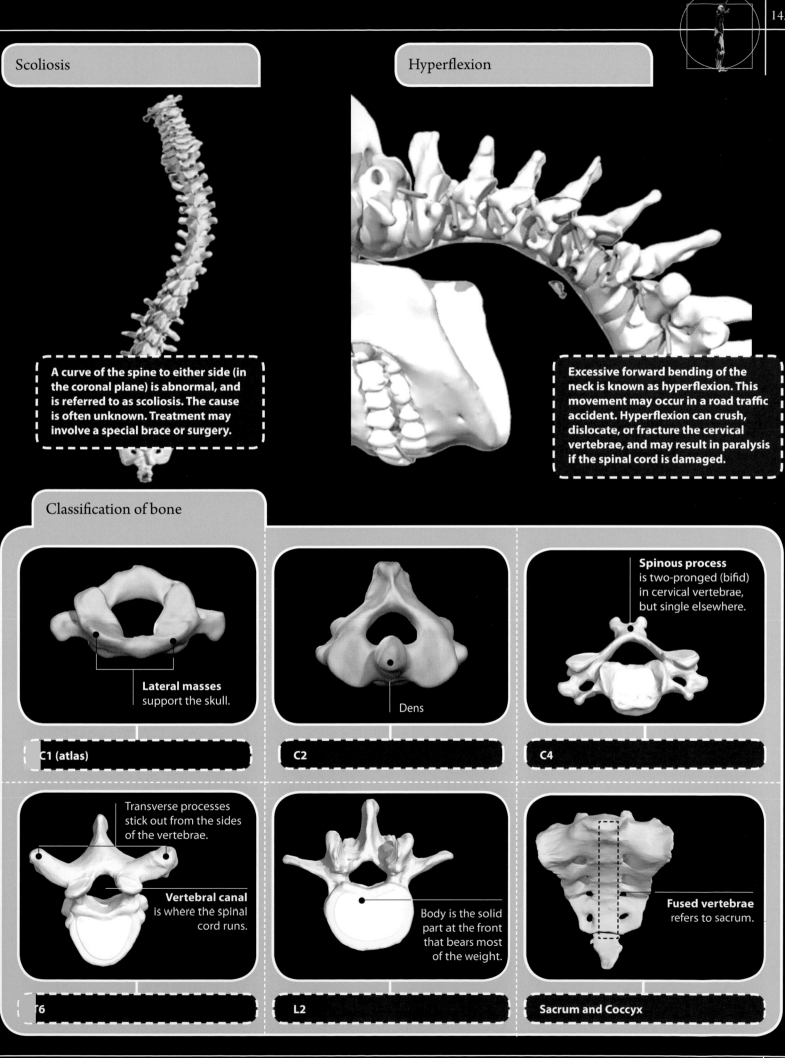

Scoliosis

A curve of the spine to either side (in the coronal plane) is abnormal, and is referred to as scoliosis. The cause is often unknown. Treatment may involve a special brace or surgery.

Hyperflexion

Excessive forward bending of the neck is known as hyperflexion. This movement may occur in a road traffic accident. Hyperflexion can crush, dislocate, or fracture the cervical vertebrae, and may result in paralysis if the spinal cord is damaged.

Classification of bone

Lateral masses support the skull.

C1 (atlas)

Dens

C2

Spinous process is two-pronged (bifid) in cervical vertebrae, but single elsewhere.

C4

Transverse processes stick out from the sides of the vertebrae.

Vertebral canal is where the spinal cord runs.

T6

Body is the solid part at the front that bears most of the weight.

L2

Fused vertebrae refers to sacrum.

Sacrum and Coccyx

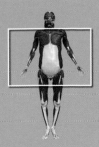

The back has a large number of muscles, running either side of the vertebral column.

Working together, they allow us to straighten our back (extension), bend to the sides (lateral flexion), and twist from side to side (rotation).

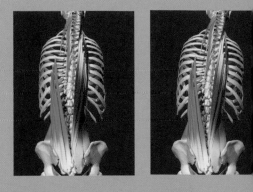

Iliocostalis cervicis
is a small muscle that extends, laterally flexes, and rotates the neck.

Spinalis capitis
extends and turns the head.

Iliocostalis thoracis
causes extension when the muscles on both sides contract. If only one side contracts, then it causes lateral flexion.

Spinalis thoracis
is a muscle that is missing in some people.

Longissimus thoracis
is the largest of the erector spinae muscles.

Hip bone
provides attachment for the erector spinae group of muscles.

Sacrum

Erector spinae

is a large group of muscles that run the length of the vertebral column. It is divided into iliocostalis, spinalis, and longissimus muscles. Each of these is then further divided according to whether it acts on the head (capitis), neck (cervicis), ribcage (thoracis), or lower back (lumborum).

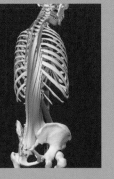

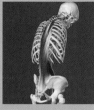
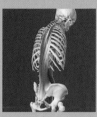
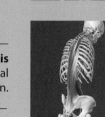
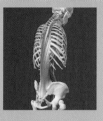
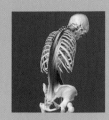
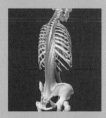
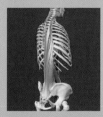
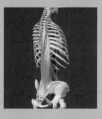
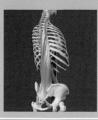

Semispinalis capitis
helps extend and turn the head.

Semispinalis thoracis
extends and rotates the vertebral
column in the thoracic region.

Supraspinous ligament
runs the length of the vertebral
column, connecting the tips of
the spinous processes of the
individual vertebrae.

**Spinalis thoracis
(erector spinae)**

Ribs

Intercostal muscles
lie between the ribs.

Multifidus
are the deepest of
the back muscles.

4th lumbar vertebra (L4)

Quadratus lumborum
is a back muscle that is not part
of either the erector spinae or
transversospinalis groups. It helps
extend and laterally flex the trunk.

Transversospinalis
is the other large group of
back muscles. It is made
up of the semispinalis and
multifidus muscles.

The spinal cord relays nerve signals between the brain and the rest of the body. It joins the brainstem as it enters the skull. The spinal cord runs inside the vertebral canal, and ends at the level of the second lumbar vertebra (L2). Below this level, the spinal nerves continue as a collection of fibers called the cauda equina.

There are thirty-one pairs of spinal nerves that supply the body. They exit the vertebral canal through gaps between the vertebrae. Like the brain, the spinal cord is covered by the meninges, and bathed in cerebrospinal fluid (CSF).

Did you know?

In diseases like meningitis, it is important for doctors to get samples of a patient's cerebrospinal fluid (CSF). This can be done by inserting a needle into the vertebral canal in a procedure called a lumbar puncture, or spinal tap. By inserting the needle well below the second lumbar vertebra (L2), this ensures that the spinal cord is not damaged by the needle.

Cervical enlargement
is a widening of the spinal cord in the neck, where the nerves that supply the arms enter and leave.

Spinal cord

Spinal nerves
emerge along the length of the spinal cord.

Lumbar enlargement
is a widening of the spinal cord in the lower back, where the nerves that supply the legs enter and leave.

Second lumbar vertebra (L2)
is where the spinal cord ends.

Cauda equina
is a collection of spinal nerves that look like a horse's tail.

Sacrum

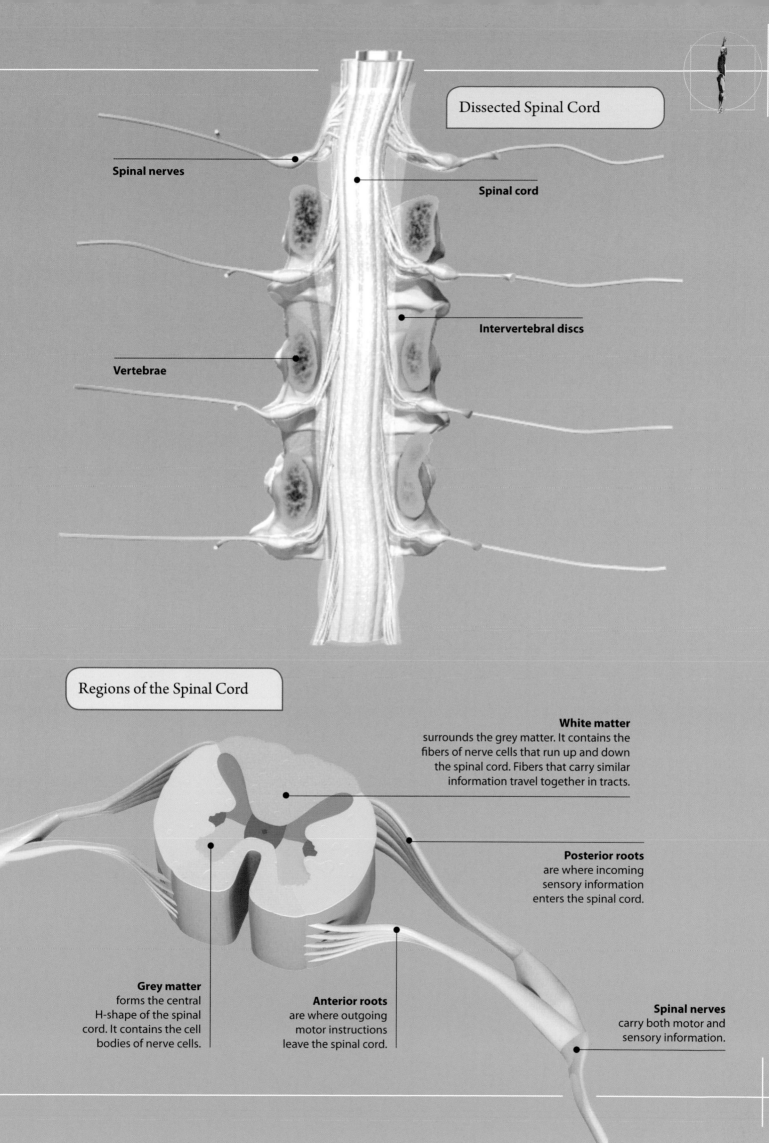

Dissected Spinal Cord

Spinal nerves

Spinal cord

Intervertebral discs

Vertebrae

Regions of the Spinal Cord

White matter
surrounds the grey matter. It contains the
fibers of nerve cells that run up and down
the spinal cord. Fibers that carry similar
information travel together in tracts.

Posterior roots
are where incoming
sensory information
enters the spinal cord.

Grey matter
forms the central
H-shape of the spinal
cord. It contains the cell
bodies of nerve cells.

Anterior roots
are where outgoing
motor instructions
leave the spinal cord.

Spinal nerves
carry both motor and
sensory information.

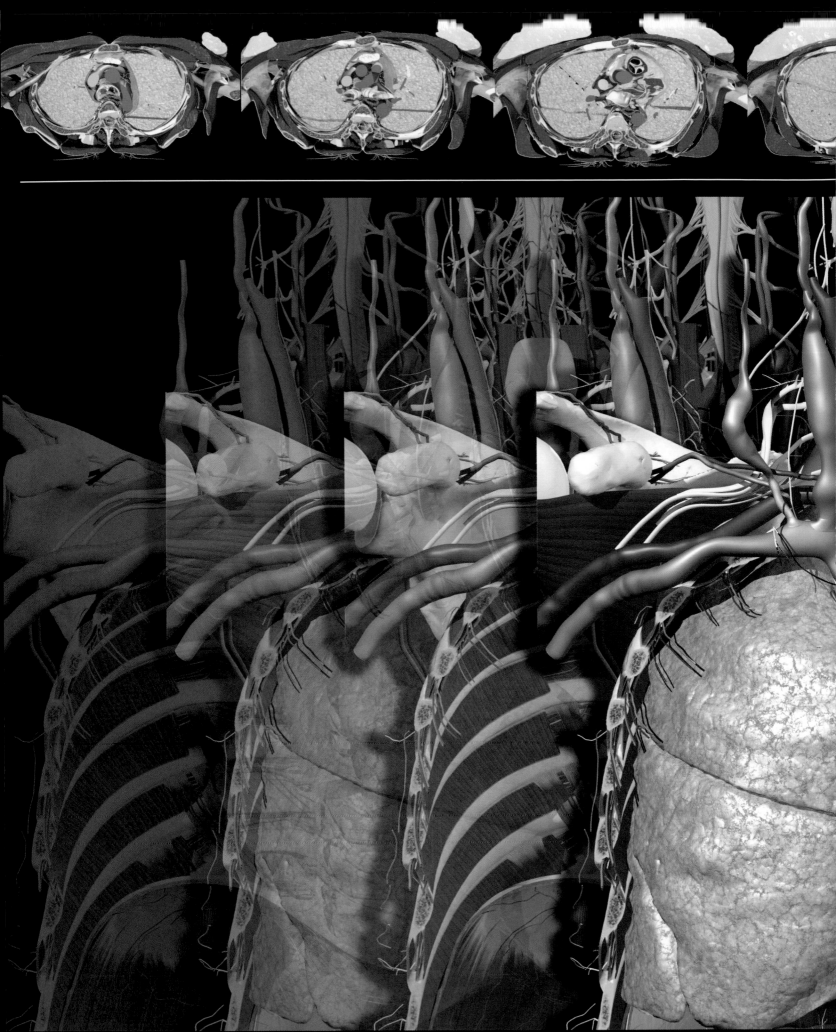

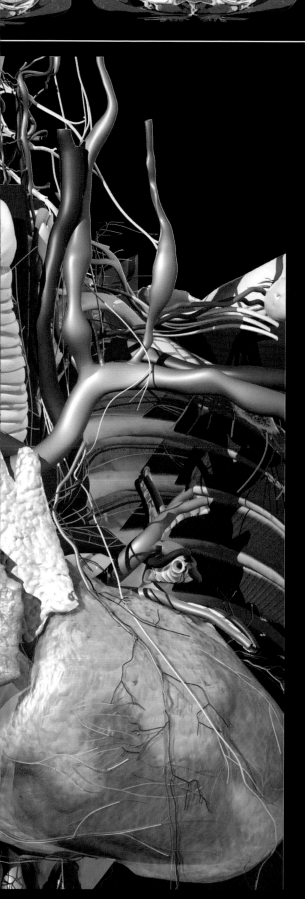

The upper part of the trunk is called the thorax, or chest. It lies between the neck and abdomen, and is mainly formed by the bony ribcage. This structure protects the contents of the thoracic cavity (the space inside the ribcage), as well as moving to allow us to breathe.

The thorax contains the heart, lungs, and large blood vessels, along with other structures moving between the neck above, and the abdomen below. The thoracic cavity is separated from the abdominal cavity by the muscular diaphragm. Numerous muscles that assist with breathing and movements of the arms and trunk are attached to the rib cage. The breasts are found at the front of the thorax.

The ribs, sternum, and thoracic vertebrae form the bony framework of the thorax, the thoracic cage. There are twelve pairs of ribs. They form joints at the back with the thoracic vertebrae. At the front, most are connected to the sternum by flexible costal cartilages.

The rib cage is mobile, with the ribs able to move up and down. This movement expands the thoracic cavity and allows us to breathe. A typical rib has a head, neck, tubercle, and body.

Thoracic Cage Anterior View

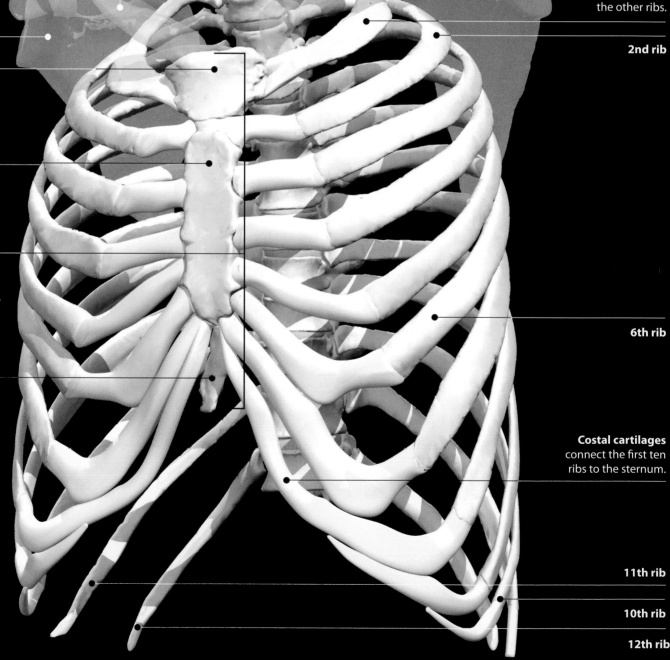

Clavicle
or collarbone, connects
to the manubrium.

Scapula
(shoulder blade).

Manubrium
is the uppermost
part of the sternum.

Body
is the large middle part
of the sternum, that
protects the heart.

Sternum
or breastbone, is the
solid central bone at the
front of the chest. It is
divided into three parts:
manubrium, body, and
xiphisternum.

Xiphisternum
is the lowermost
part of the sternum.

1st rib
is broader than
the other ribs.

2nd rib

6th rib

Costal cartilages
connect the first ten
ribs to the sternum.

11th rib

10th rib

12th rib

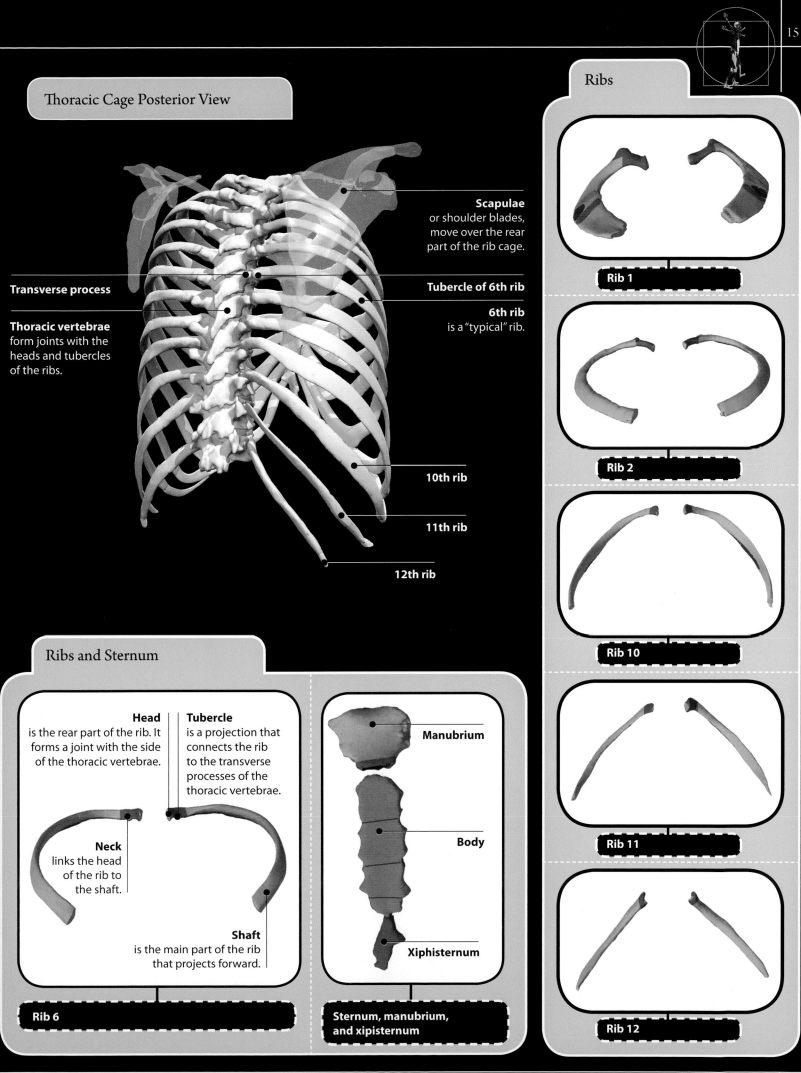

Thoracic Cage Posterior View

Scapulae or shoulder blades, move over the rear part of the rib cage.

Transverse process

Tubercle of 6th rib

6th rib is a "typical" rib.

Thoracic vertebrae form joints with the heads and tubercles of the ribs.

10th rib

11th rib

12th rib

Ribs and Sternum

Head is the rear part of the rib. It forms a joint with the side of the thoracic vertebrae.

Tubercle is a projection that connects the rib to the transverse processes of the thoracic vertebrae.

Neck links the head of the rib to the shaft.

Shaft is the main part of the rib that projects forward.

Manubrium

Body

Xiphisternum

Rib 6

Sternum, manubrium, and xipisternum

Ribs

Rib 1

Rib 2

Rib 10

Rib 11

Rib 12

MUSCLES OF THE THORAX

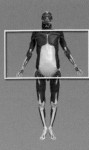

The main muscles of the thorax are the diaphragm and intercostal muscles. They are called muscles of respiration, as their main role is to help us breathe.

The intercostal muscles lie between the ribs, and move the rib cage up and down.

The diaphragm is a large dome-shaped sheet of muscle that separates the thoracic and abdominal cavities. When it contracts, it flattens and moves downward. This increases the size of the thoracic cavity and helps us to breathe in.

Other muscles that lie on the surface of the rib cage can help when we need to breathe hard and fast, for example, during exercise.

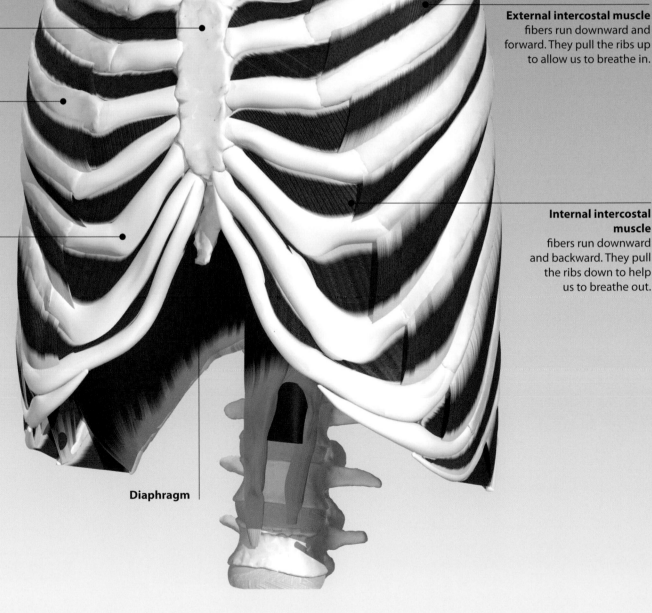

Sternum (breastbone)

Ribs
are moved up and down by
the intercostal muscles.

Costal cartilages
connect the ribs
to the sternum.

External intercostal muscle
fibers run downward and
forward. They pull the ribs up
to allow us to breathe in.

**Internal intercostal
muscle**
fibers run downward
and backward. They pull
the ribs down to help
us to breathe out.

Diaphragm

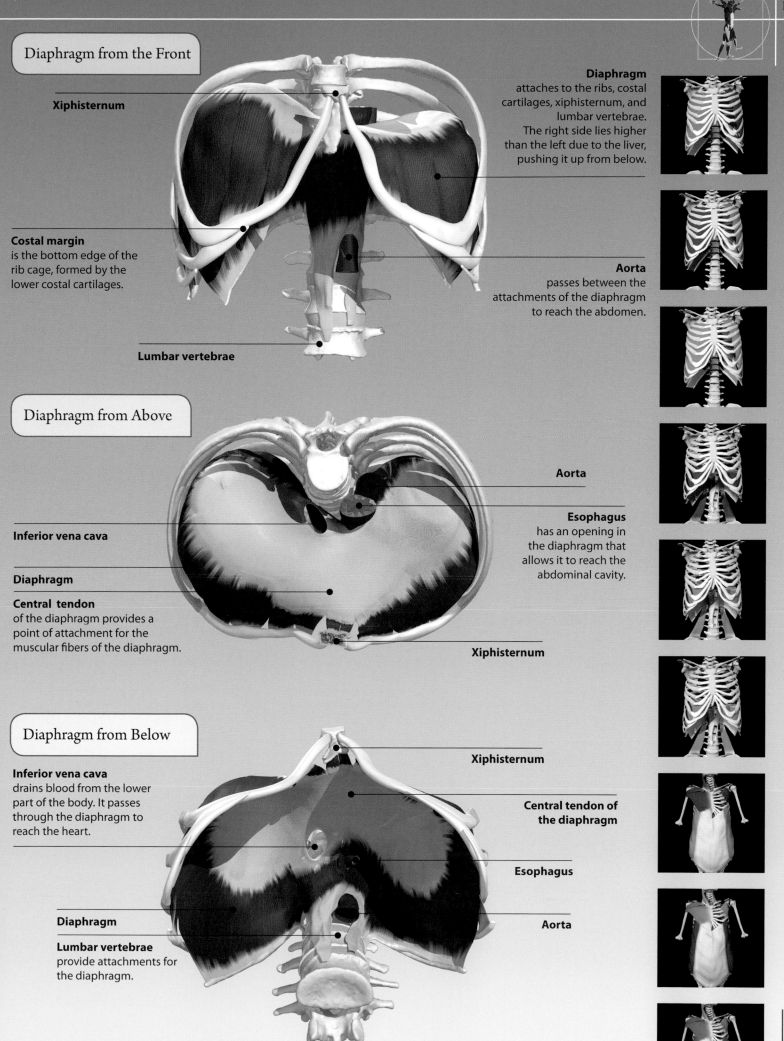

Diaphragm from the Front

Xiphisternum

Costal margin
is the bottom edge of the
rib cage, formed by the
lower costal cartilages.

Lumbar vertebrae

Diaphragm
attaches to the ribs, costal
cartilages, xiphisternum, and
lumbar vertebrae.
The right side lies higher
than the left due to the liver,
pushing it up from below.

Aorta
passes between the
attachments of the diaphragm
to reach the abdomen.

Diaphragm from Above

Inferior vena cava

Diaphragm

Central tendon
of the diaphragm provides a
point of attachment for the
muscular fibers of the diaphragm.

Aorta

Esophagus
has an opening in
the diaphragm that
allows it to reach the
abdominal cavity.

Xiphisternum

Diaphragm from Below

Inferior vena cava
drains blood from the lower
part of the body. It passes
through the diaphragm to
reach the heart.

Diaphragm

Lumbar vertebrae
provide attachments for
the diaphragm.

Xiphisternum

**Central tendon of
the diaphragm**

Esophagus

Aorta

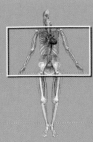

The thorax contains the large blood vessels that enter and leave the heart, along with numerous other branches.

Twelve pairs of thoracic spinal nerves innervate the skin of the thorax and intercostal muscles.

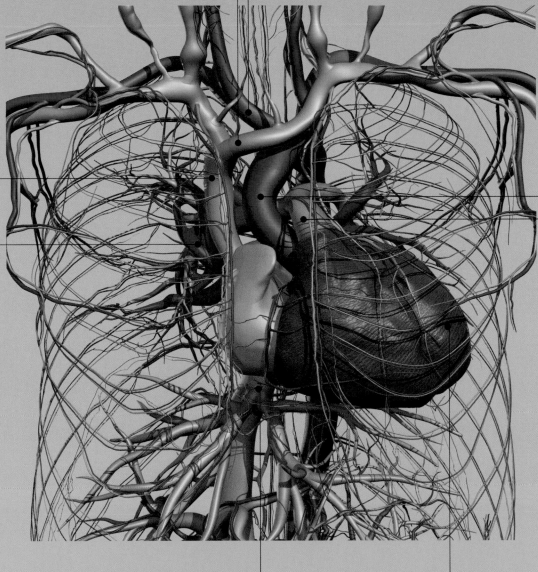

Brachiocephalic veins
drain blood from the head, neck, and arms, and unite to form the superior vena cava.

Brachiocephalic artery
is only present on the right side. It divides into the right common carotid and right subclavian arteries.

Superior vena cava
drains venous blood from the upper body to the right side of the heart.

Ascending aorta
is the first part of the large blood vessel that delivers oxygen-rich blood to the rest of the body.

Pulmonary veins
carry oxygen-rich blood from the lungs to the left side of the heart, ready to be pumped to the rest of the body.

Pulmonary trunk
delivers deoxygenated blood to the lungs for it to be resupplied with oxygen.

Inferior vena cava
drains venous blood from the lower part of the body to the right side of the heart.

Heart
pumps blood around the lungs and the body.

Arteries and Veins of the Thorax

Thoracic Aorta

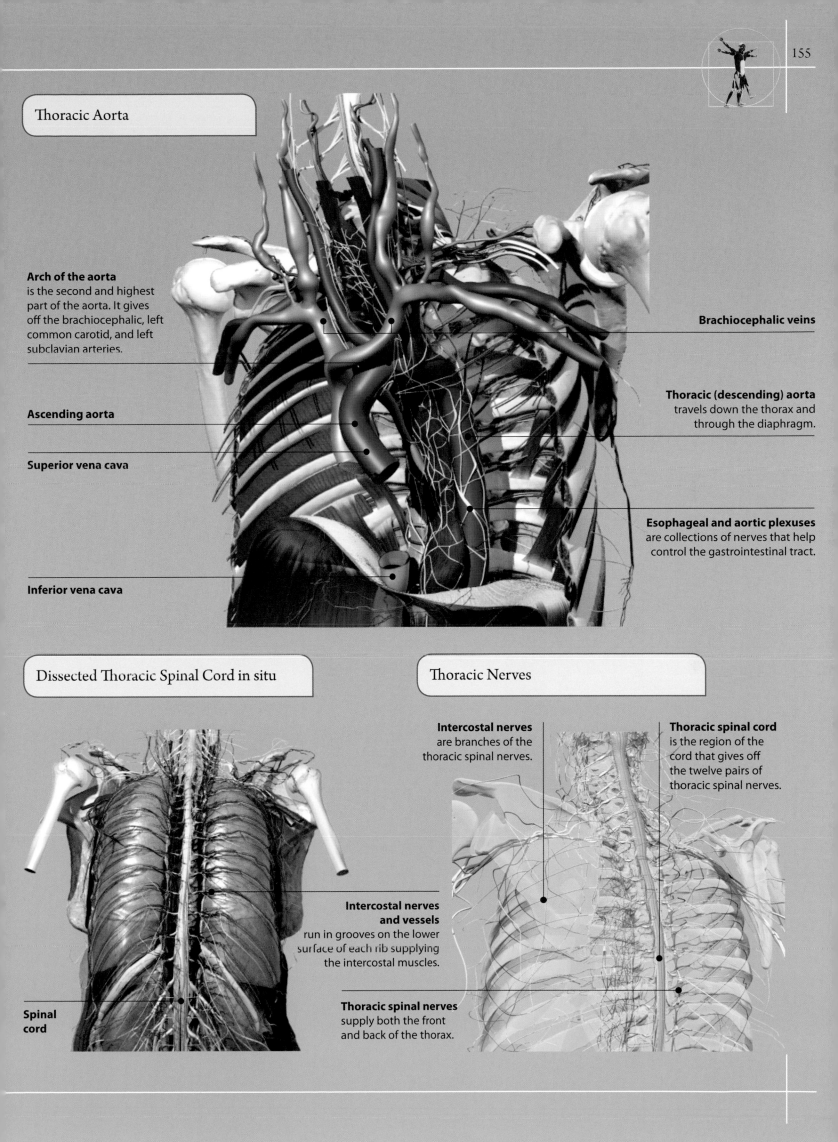

Arch of the aorta
is the second and highest part of the aorta. It gives off the brachiocephalic, left common carotid, and left subclavian arteries.

Ascending aorta

Superior vena cava

Inferior vena cava

Brachiocephalic veins

Thoracic (descending) aorta
travels down the thorax and through the diaphragm.

Esophageal and aortic plexuses
are collections of nerves that help control the gastrointestinal tract.

Dissected Thoracic Spinal Cord in situ

Intercostal nerves and vessels
run in grooves on the lower surface of each rib supplying the intercostal muscles.

Spinal cord

Thoracic Nerves

Intercostal nerves
are branches of the thoracic spinal nerves.

Thoracic spinal cord
is the region of the cord that gives off the twelve pairs of thoracic spinal nerves.

Thoracic spinal nerves
supply both the front and back of the thorax.

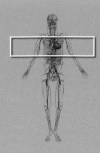

The heart is a fist-sized, muscular organ. It is located in the chest between the lungs, and is tilted to the left side. It continuously pumps blood around the lungs and body. This provides the cells and tissues of the body with the oxygen and nutrients they need to stay alive.

The heart is divided into right and left sides, which pump blood around different circuits. The right side of the heart receives deoxygenated blood from the body and pumps this to the lungs. The left side of the heart receives oxygenated blood from the lungs and pumps this to the rest of the body. Each side of the heart has two chambers: an atrium (priming chamber) and a ventricle (pumping chamber).

Aorta
delivers oxygenated blood to the body.

Superior vena cava
delivers deoxygenated blood from the upper body into the right atrium.

Coronary vessels
supply the heart with oxygen and nutrients to allow it to keep pumping.

Cardiac muscle
is a specialized tissue that is able to beat continuously without tiring. It generates and conducts its own signals to contract, coordinated by a built-in pacemaker.

Right atrium
receives deoxygenated venous blood from the body.

Apex
is the tip of the heart. It is this part that can be felt beating on the left side of the chest wall.

Right ventricle
is the chamber closest to the front of the heart.

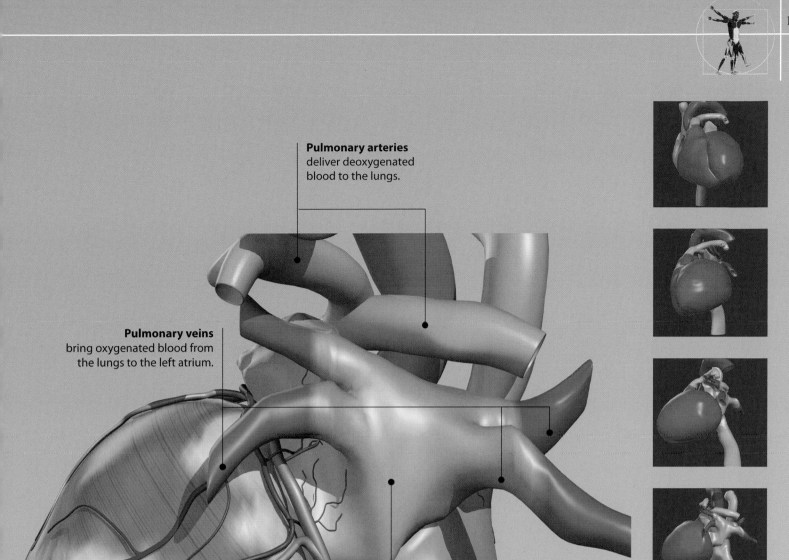

Pulmonary arteries deliver deoxygenated blood to the lungs.

Pulmonary veins bring oxygenated blood from the lungs to the left atrium.

Left ventricle pumps blood to the body.

Left atrium is the chamber at the back of the heart.

Inferior vena cava delivers deoxygenated blood from the lower body to the right atrium.

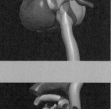

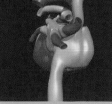

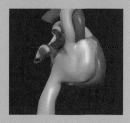

The heart has four chambers (two atria and two ventricles) in which blood is collected and pumped. The chambers of the right and left side of the heart are separated by a fibromuscular wall called the septum.

Atrioventricular valves lie between the atria and ventricles on each side. These ensure that blood only flows in one direction through the heart.

Did you know?

On average, the human heart beats about 100,000 times a day; that's approximately 35 million times every year without a rest!

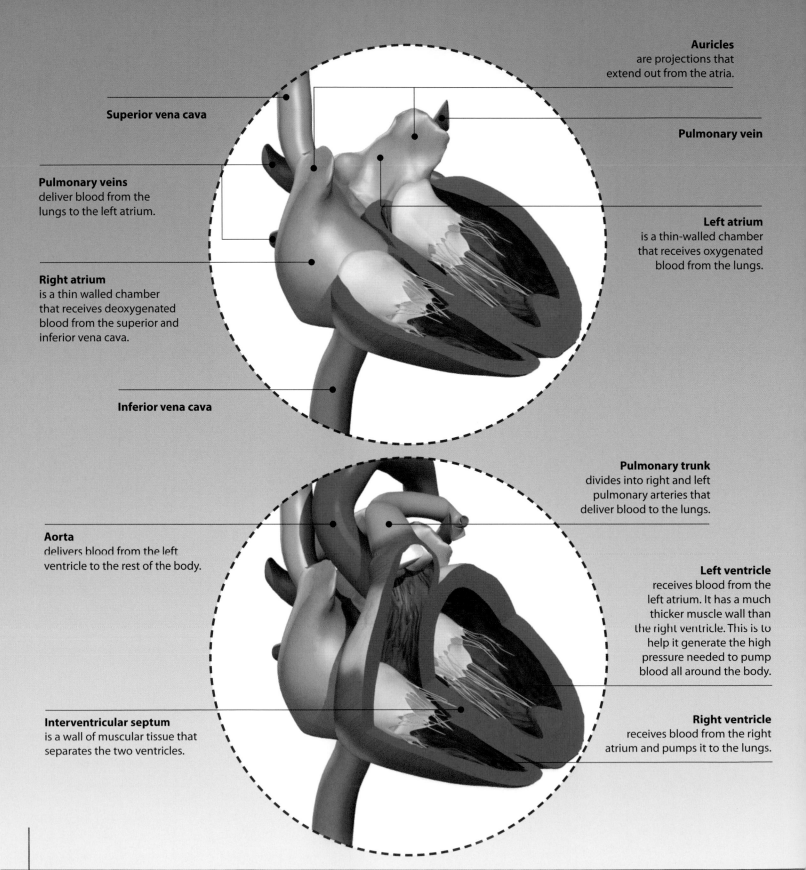

Auricles
are projections that extend out from the atria.

Superior vena cava

Pulmonary vein

Pulmonary veins
deliver blood from the lungs to the left atrium.

Left atrium
is a thin-walled chamber that receives oxygenated blood from the lungs.

Right atrium
is a thin walled chamber that receives deoxygenated blood from the superior and inferior vena cava.

Inferior vena cava

Pulmonary trunk
divides into right and left pulmonary arteries that deliver blood to the lungs.

Aorta
delivers blood from the left ventricle to the rest of the body.

Left ventricle
receives blood from the left atrium. It has a much thicker muscle wall than the right ventricle. This is to help it generate the high pressure needed to pump blood all around the body.

Interventricular septum
is a wall of muscular tissue that separates the two ventricles.

Right ventricle
receives blood from the right atrium and pumps it to the lungs.

Atrioventricular valves
stop blood in the
ventricles flowing
back into the atria.

Right ventricle

Papillary muscles
are attached to the edges of the
atrioventricular valves via the
chordae tendinae. They prevent
the valves from turning inside
out and leaking.

Left ventricle

Chordae tendinae
are long, thin fibers
attached to the edge of
the atrioventricular valves.

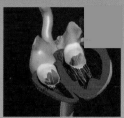

Interventricular sulcus
is a groove on the
surface of the heart.
It is filled with fatty
tissue, and marks the
division of the right
and left ventricles.

Right ventricle

Left ventricle

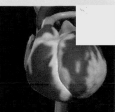

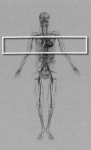

The heart is found within the thorax between the two lungs. It sits on top of the diaphragm, behind the sternum (breastbone) and ribs. It is shaped like a cone, with an apex that points to the front and left, and a base pointing backward and to the right.

The heart is surrounded and contained by a sac called the pericardium. This has a smooth inner serous layer, and a tough outer fibrous layer.

Outer Layer of Pericardium

Fibrous pericardium is the tough outer layer. It is attached to the upper surface of the diaphragm. It prevents the heart from expanding or moving from its position.

Diaphragm

Inner Layer of Pericardium

Serous pericardium consists of two thin, smooth layers, that move over each other. They produce pericardial fluid that reduces friction as the heart moves with each beat.

Diaphragm

Position of Pericardium

Esophagus　**Aorta**

Inferior vena cava
enters the heart
almost as soon as it
passes through the
diaphragm.

Pericardium
is attached to the
upper surface of
the diaphragm.

Diaphragm

Heart Within the Thorax

Superior vena cava

Base
of the heart is where
the large blood
vessels enter and
leave the heart.

Aorta

Pulmonary trunk

Ribs

Apex
is on the left side of the
chest in most people.

Abdominal aorta
delivers blood to the
lower half of the body.

Inferior vena cava

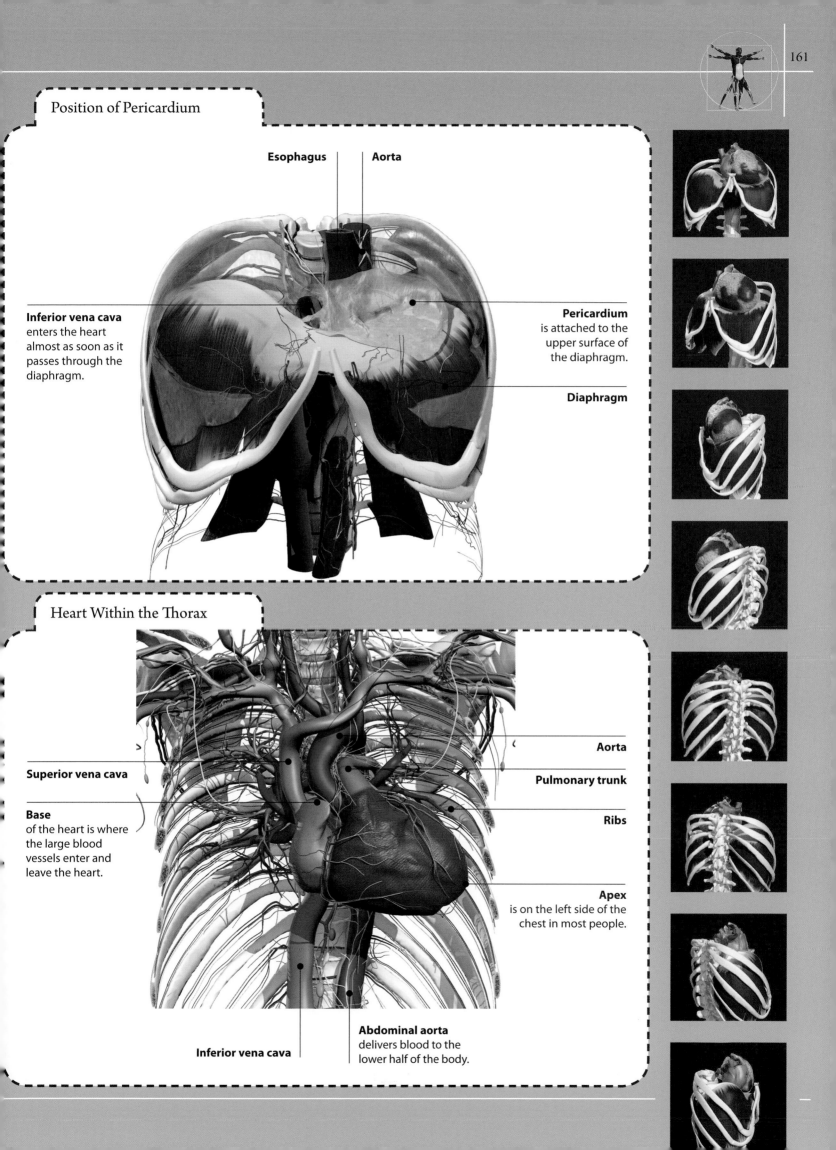

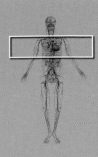

The heart has four valves. Their function is to allow blood to flow in only one direction through the chambers of the heart. They open and close in response to pressure changes within the chambers of the heart.

The atrioventricular valves lie between the atria and ventricles. The semilunar valves lie between the ventricles and large blood vessels carrying blood away from the heart. The valves are attached to the fibrous tissue that forms the framework of the heart.

It is the closing of the atrioventricular and semilunar heart valves that is heard through a stethoscope as the heart sounds "LUB" and "DUB."

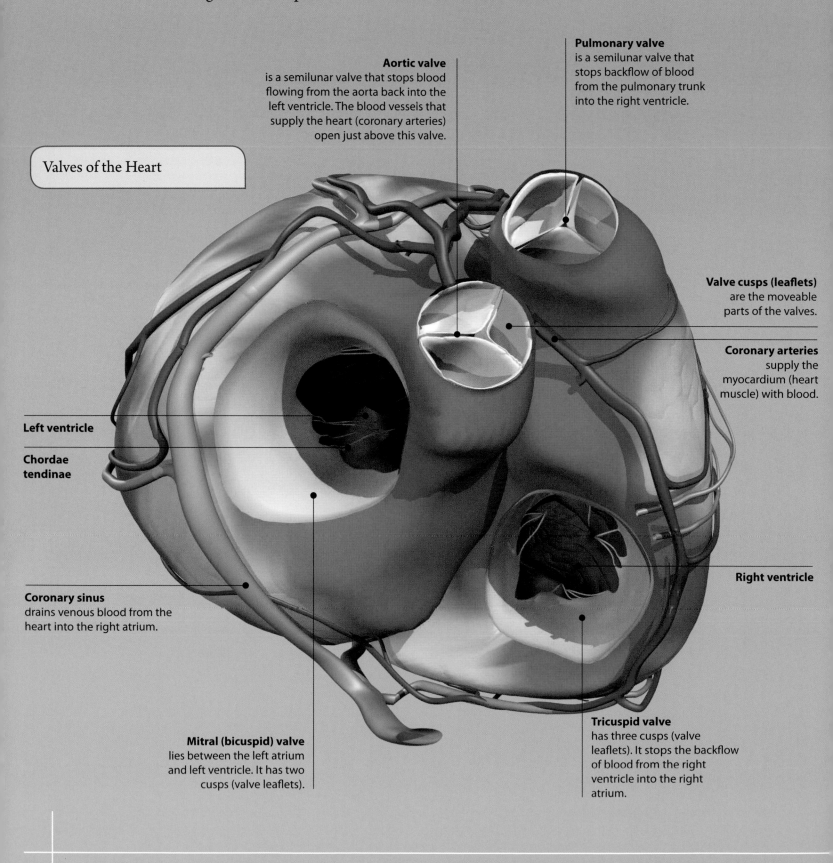

Valves of the Heart

Aortic valve
is a semilunar valve that stops blood flowing from the aorta back into the left ventricle. The blood vessels that supply the heart (coronary arteries) open just above this valve.

Pulmonary valve
is a semilunar valve that stops backflow of blood from the pulmonary trunk into the right ventricle.

Valve cusps (leaflets)
are the moveable parts of the valves.

Coronary arteries
supply the myocardium (heart muscle) with blood.

Left ventricle

Chordae tendinae

Right ventricle

Coronary sinus
drains venous blood from the heart into the right atrium.

Mitral (bicuspid) valve
lies between the left atrium and left ventricle. It has two cusps (valve leaflets).

Tricuspid valve
has three cusps (valve leaflets). It stops the backflow of blood from the right ventricle into the right atrium.

Cross Section Through the Heart

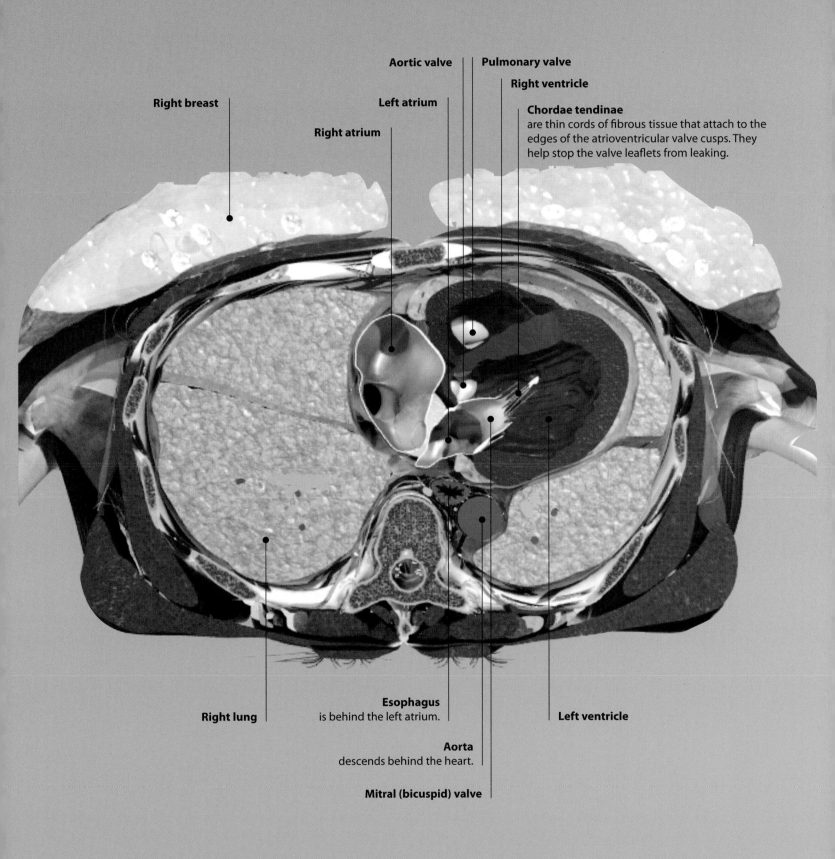

Aortic valve

Pulmonary valve

Right ventricle

Left atrium

Right breast

Chordae tendinae
are thin cords of fibrous tissue that attach to the edges of the atrioventricular valve cusps. They help stop the valve leaflets from leaking.

Right atrium

Esophagus
is behind the left atrium.

Right lung

Left ventricle

Aorta
descends behind the heart.

Mitral (bicuspid) valve

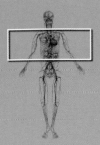

The myocardium (heart muscle) needs a constant supply of oxygen- and nutrient-rich blood to keep it pumping. This is delivered by the right and left coronary arteries. They branch off the aorta, immediately after the aortic valve. Between them they supply the entire heart.

Any interruption to this crucial blood supply can stop the heart from pumping properly, and may even lead to death.

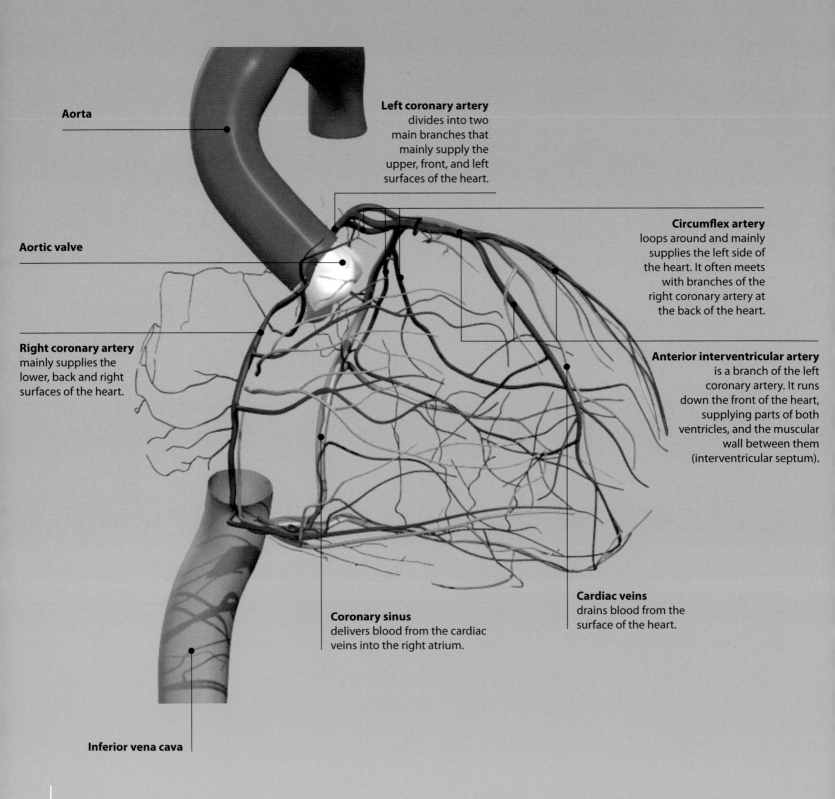

Aorta

Left coronary artery
divides into two main branches that mainly supply the upper, front, and left surfaces of the heart.

Aortic valve

Circumflex artery
loops around and mainly supplies the left side of the heart. It often meets with branches of the right coronary artery at the back of the heart.

Right coronary artery
mainly supplies the lower, back and right surfaces of the heart.

Anterior interventricular artery
is a branch of the left coronary artery. It runs down the front of the heart, supplying parts of both ventricles, and the muscular wall between them (interventricular septum).

Cardiac veins
drains blood from the surface of the heart.

Coronary sinus
delivers blood from the cardiac veins into the right atrium.

Inferior vena cava

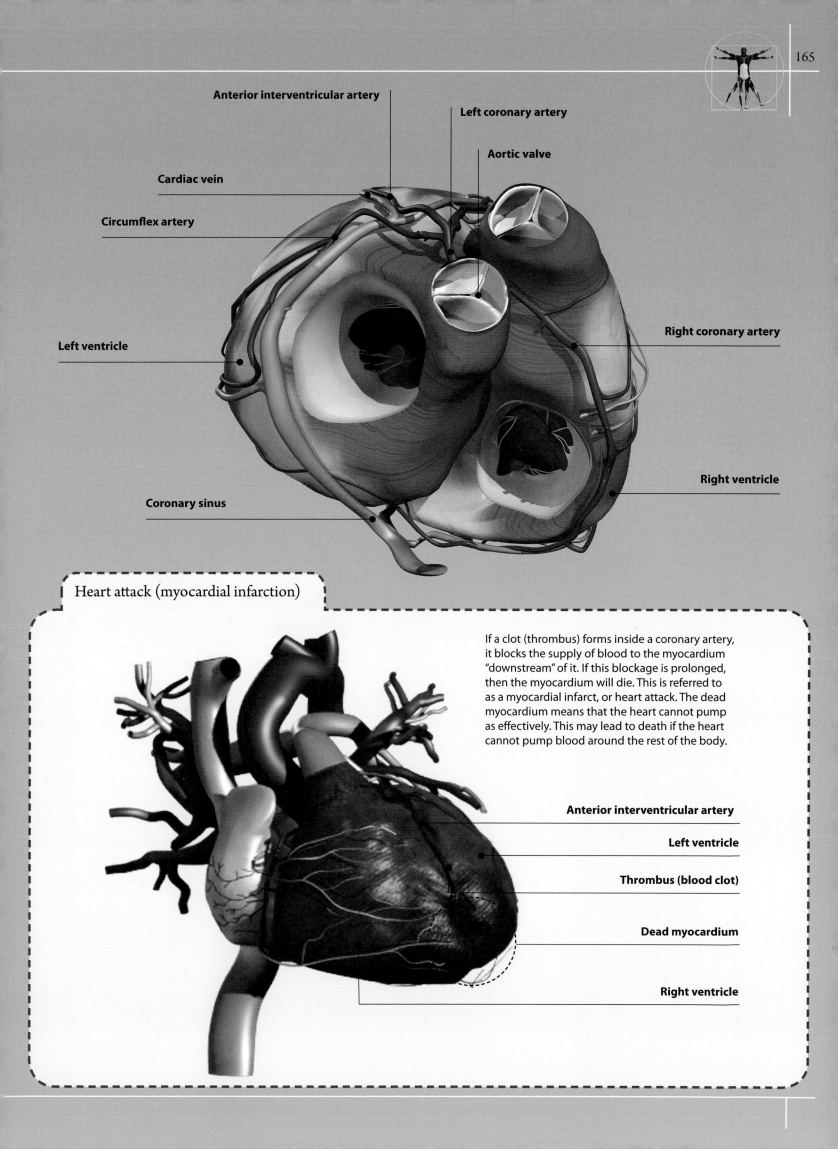

Anterior interventricular artery

Left coronary artery

Aortic valve

Cardiac vein

Circumflex artery

Right coronary artery

Left ventricle

Right ventricle

Coronary sinus

Heart attack (myocardial infarction)

If a clot (thrombus) forms inside a coronary artery, it blocks the supply of blood to the myocardium "downstream" of it. If this blockage is prolonged, then the myocardium will die. This is referred to as a myocardial infarct, or heart attack. The dead myocardium means that the heart cannot pump as effectively. This may lead to death if the heart cannot pump blood around the rest of the body.

Anterior interventricular artery

Left ventricle

Thrombus (blood clot)

Dead myocardium

Right ventricle

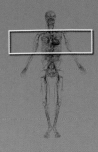

The heart has its own rhythmic "beat" generator located in the right atrium. This sends regular electrical signals, via specialized rapidly conducting cells, to the rest of the heart. These electrical signals instruct the cardiac muscle cells (cardiomyocytes) of the atria and ventricles to contract, causing the heart to beat and pump blood around the body. The rate at which signals are generated can be varied, and depends upon the needs of the body, i.e. increased in exercise, and decreased at rest.

1

The electrical and conducting system of the heart is at rest, ready to start another beat.

2

An electrical signal is generated in the sinoatrial node (SAN), the pacemaker of the heart.

3

The signal spreads from the SAN, through the right and left atrium. . .

4

. . .via the atrial conducting pathways. This signal tells the atrial cardiomyocytes to contract.

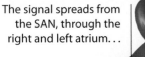

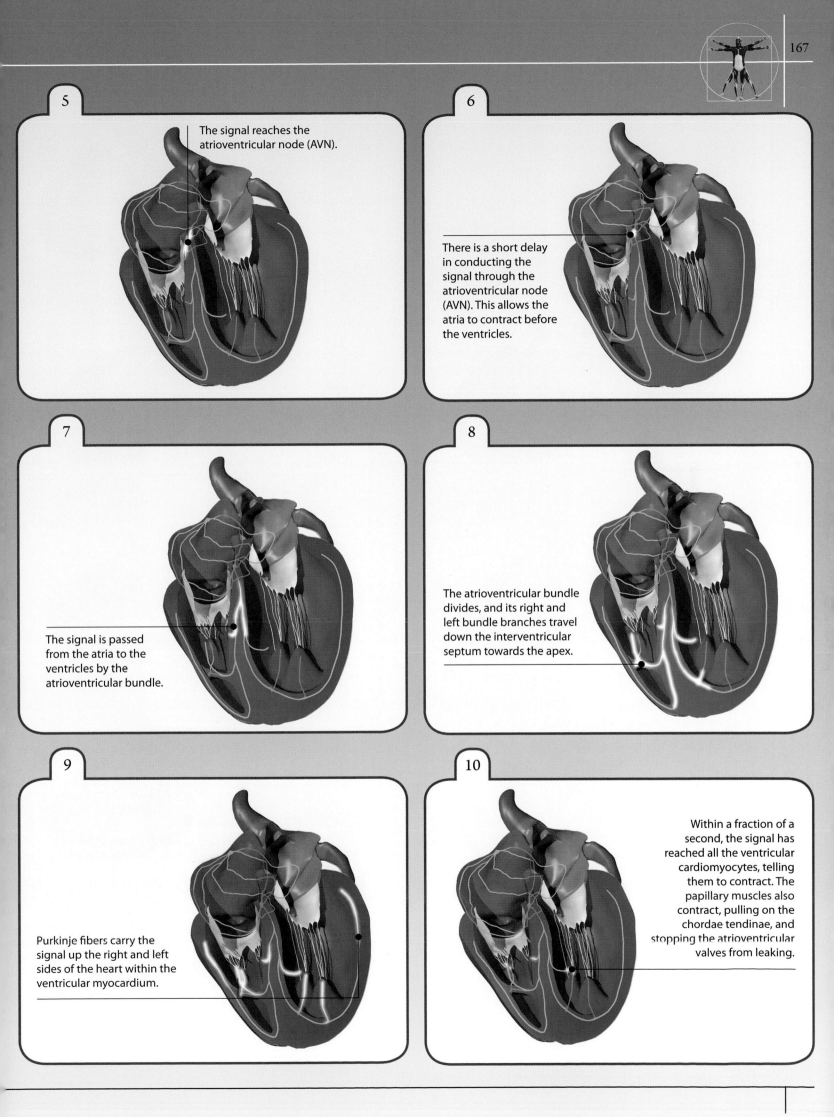

5 The signal reaches the atrioventricular node (AVN).

6 There is a short delay in conducting the signal through the atrioventricular node (AVN). This allows the atria to contract before the ventricles.

7 The signal is passed from the atria to the ventricles by the atrioventricular bundle.

8 The atrioventricular bundle divides, and its right and left bundle branches travel down the interventricular septum towards the apex.

9 Purkinje fibers carry the signal up the right and left sides of the heart within the ventricular myocardium.

10 Within a fraction of a second, the signal has reached all the ventricular cardiomyocytes, telling them to contract. The papillary muscles also contract, pulling on the chordae tendinae, and stopping the atrioventricular valves from leaking.

The cardiac cycle is the ordered sequence of events that take place with each beat of the heart. Events happen on the right and left sides of the heart at the same time.

The right side of the heart pumps deoxygenated blood to the lungs, around the pulmonary circuit. The left side of the heart pumps oxygenated blood to the rest of the body, around the systemic circuit. Contraction of a heart chamber is called systole. Relaxation of a heart chamber is called diastole.

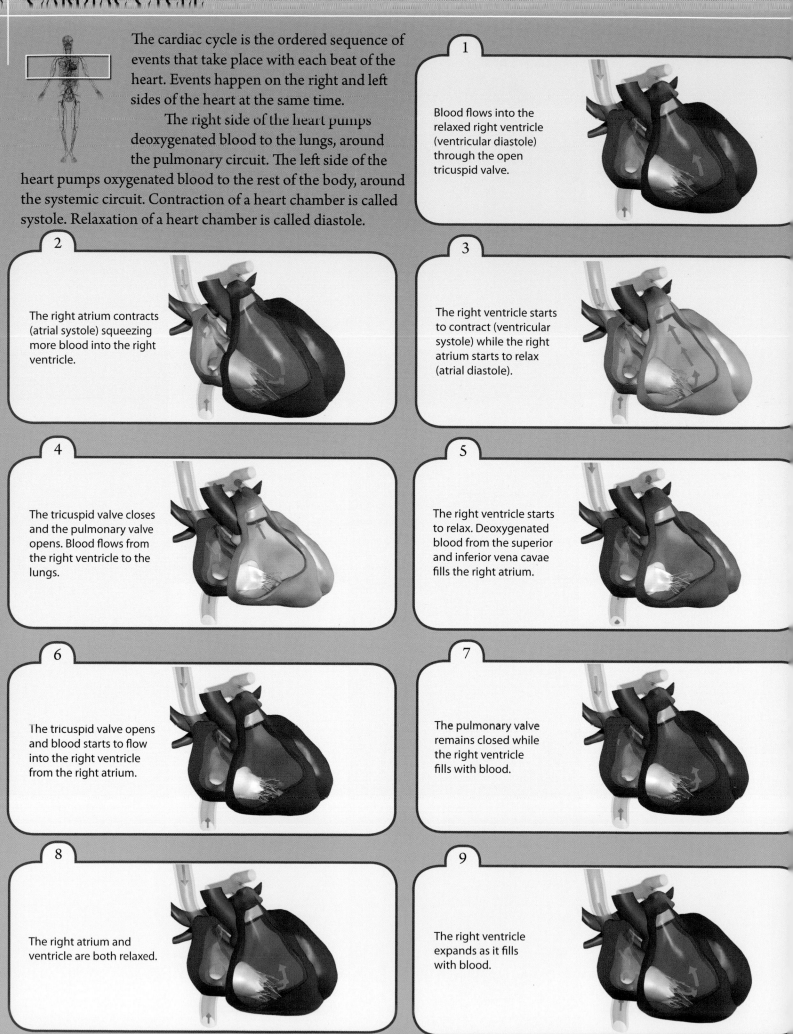

1
Blood flows into the relaxed right ventricle (ventricular diastole) through the open tricuspid valve.

2
The right atrium contracts (atrial systole) squeezing more blood into the right ventricle.

3
The right ventricle starts to contract (ventricular systole) while the right atrium starts to relax (atrial diastole).

4
The tricuspid valve closes and the pulmonary valve opens. Blood flows from the right ventricle to the lungs.

5
The right ventricle starts to relax. Deoxygenated blood from the superior and inferior vena cavae fills the right atrium.

6
The tricuspid valve opens and blood starts to flow into the right ventricle from the right atrium.

7
The pulmonary valve remains closed while the right ventricle fills with blood.

8
The right atrium and ventricle are both relaxed.

9
The right ventricle expands as it fills with blood.

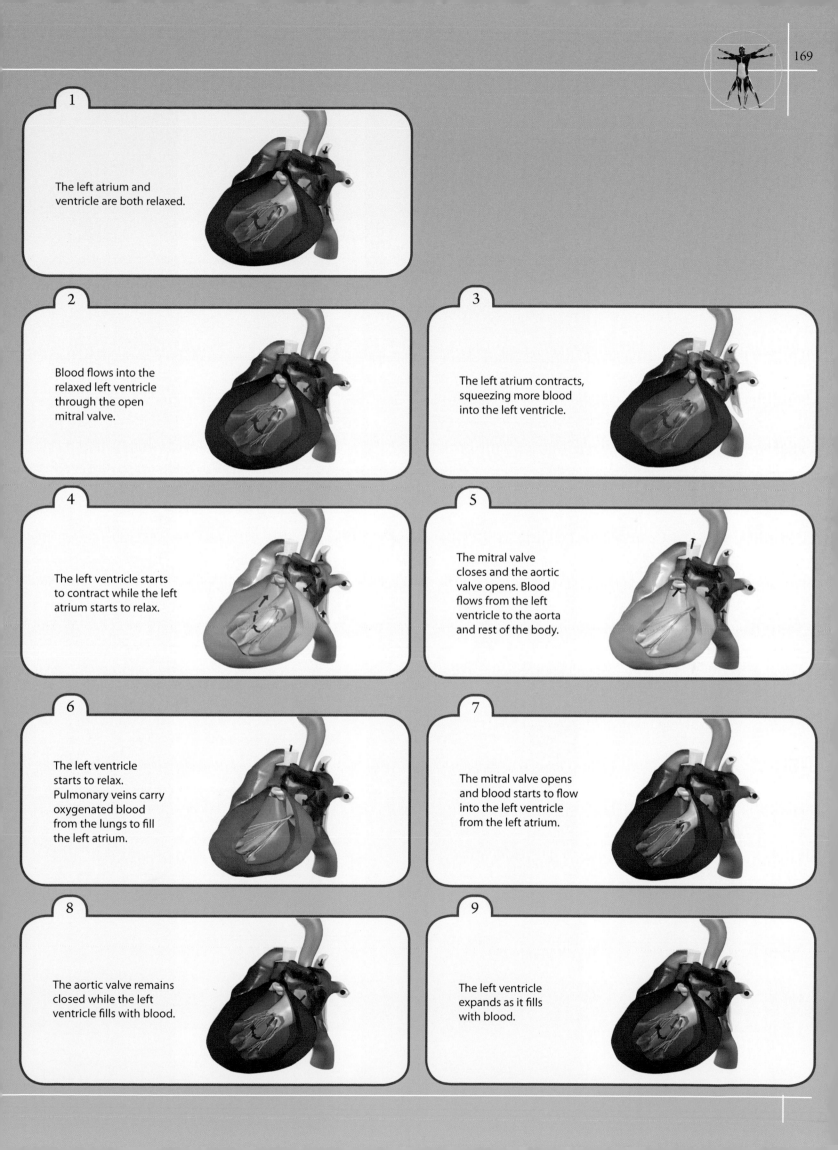

1

The left atrium and ventricle are both relaxed.

2

Blood flows into the relaxed left ventricle through the open mitral valve.

3

The left atrium contracts, squeezing more blood into the left ventricle.

4

The left ventricle starts to contract while the left atrium starts to relax.

5

The mitral valve closes and the aortic valve opens. Blood flows from the left ventricle to the aorta and rest of the body.

6

The left ventricle starts to relax. Pulmonary veins carry oxygenated blood from the lungs to fill the left atrium.

7

The mitral valve opens and blood starts to flow into the left ventricle from the left atrium.

8

The aortic valve remains closed while the left ventricle fills with blood.

9

The left ventricle expands as it fills with blood.

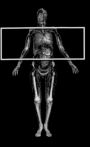

The lungs are contained within our rib cage. They are like balloons that inflate and deflate with each breath, and can hold around 12 pints of air. There are numerous air passages in the lung that repeatedly branch and divide, starting with the trachea and bronchi, and ending eventually at the alveoli.

Trachea
or windpipe is the main air passage that connects the larynx to the main bronchi. Rings of cartilage reinforce the walls of the trachea.

Cricoid cartilage
is the lowest part of the larynx (voice box).

Main bronchi
The trachea divides into two main bronchi which supply the left and right lungs.

Lobar bronchi
branch off from main bronchi and each one supplies a different lobe of the lung.

Segmental bronchi
branch off the lobar bronchi and supply a segment of the lung.

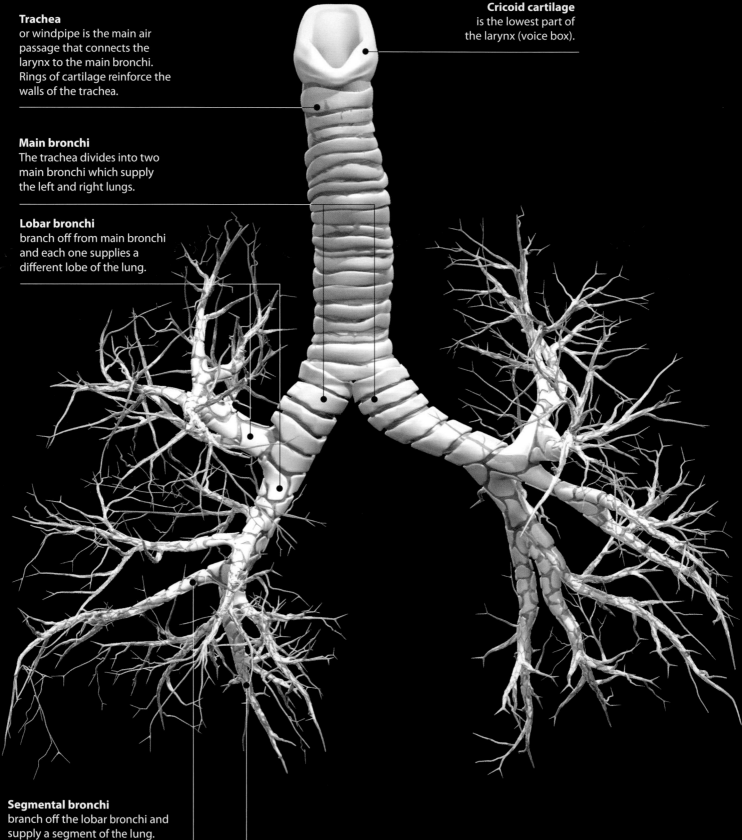

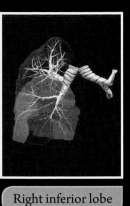

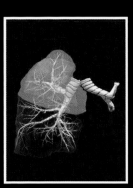

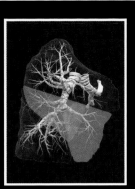

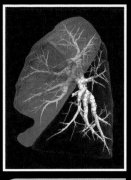

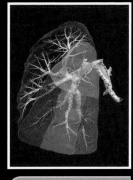

| Right inferior lobe | Right superior lobe | Right middle lobe | Left superior lobe | Left inferior lobe |

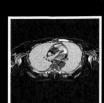

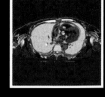

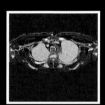

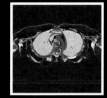

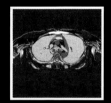

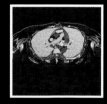

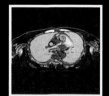

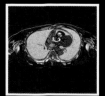

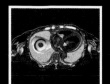

Lobes

Each lung is divided into lobes. There are three lobes in the right lung (upper, middle, and lower), and two lobes in the left lung (upper and lower). The lobes are further divided into segments.

Fissures

are narrow grooves on the lung surface that divide the lung into lobes.

Trachea

Heart

Pleura

The lungs are contained within a thin membranous sac called the pleura. The pleura covers the surface of the lung as well as lining the inside of the rib cage. It produces a small amount of fluid, which acts like a lubricant, allowing the lungs to move freely within the rib cage every time we breathe.

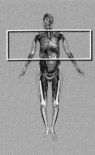

After dividing and branching up to twenty-five times, the bronchial tree eventually forms the terminal bronchioles, which deliver air to the alveoli. Within each of the individual alveoli, oxygen from the air is transferred into the blood, and excess carbon dioxide removed from it, in a process called gas exchange. This takes place across a thin barrier called the respiratory membrane.

Terminal bronchioles
deliver air to the alveoli where gas exchange can occur.

Alveoli
are round, blind-ended sacs where gas exchange takes place.

Respiratory membrane
is the thin barrier that separates the gases in the alveoli from the blood in the pulmonary capillaries.

Pulmonary capillaries
form a network of blood vessels surrounding the alveoli. The blood arriving at the alveoli is low in oxygen. Following gas transfer, the blood leaving the alveoli is rich in oxygen, and returns to the heart to be pumped to the rest of the body.

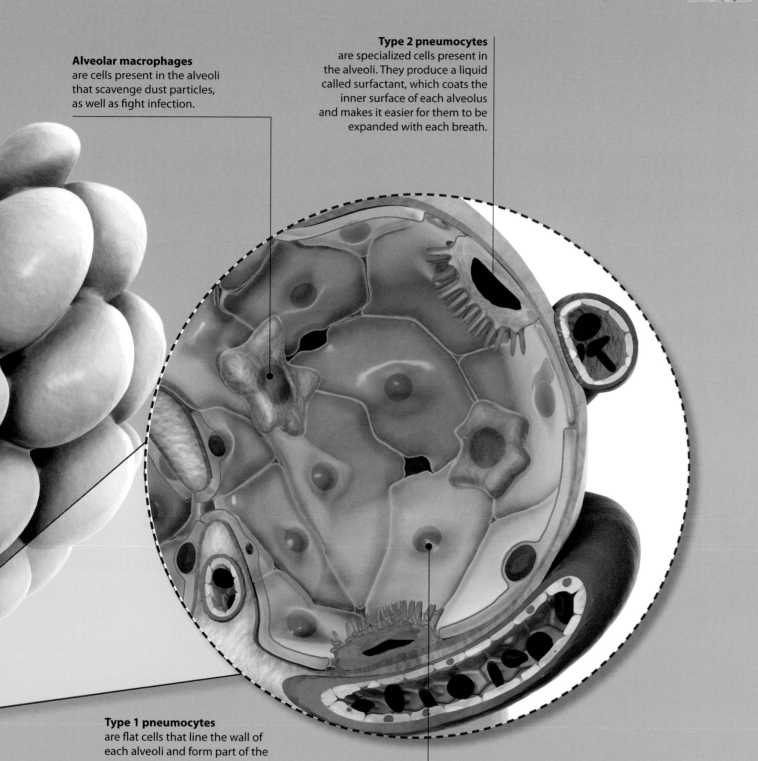

Alveolar macrophages
are cells present in the alveoli
that scavenge dust particles,
as well as fight infection.

Type 2 pneumocytes
are specialized cells present in
the alveoli. They produce a liquid
called surfactant, which coats the
inner surface of each alveolus
and makes it easier for them to be
expanded with each breath.

Type 1 pneumocytes
are flat cells that line the wall of
each alveoli and form part of the
respiratory membrane.

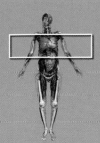

The lungs are lined by a thin, double-layered membranous sac called the pleura. In most places the two pleural layers are next to each other, with a small amount of fluid between them. This pleural fluid lubricates the movements of the lungs during breathing.

The space between the two lungs is called the mediastinum. Its contents include the heart, large blood vessels, and esophagus.

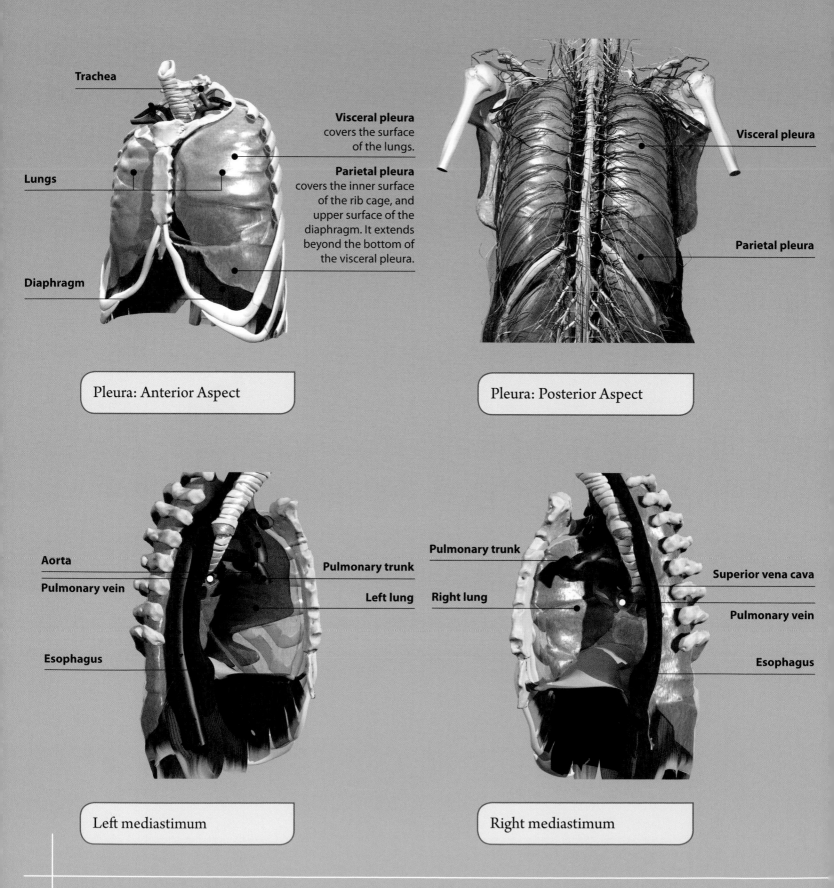

Trachea

Lungs

Diaphragm

Visceral pleura
covers the surface
of the lungs.

Parietal pleura
covers the inner surface
of the rib cage, and
upper surface of the
diaphragm. It extends
beyond the bottom of
the visceral pleura.

Pleura: Anterior Aspect

Visceral pleura

Parietal pleura

Pleura: Posterior Aspect

Aorta

Pulmonary vein

Esophagus

Pulmonary trunk

Left lung

Left mediastimum

Pulmonary trunk

Right lung

Superior vena cava

Pulmonary vein

Esophagus

Right mediastimum

Pulmonary Vessels

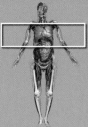

The two pulmonary arteries deliver deoxygenated blood to the lungs from the right ventricle. The four pulmonary veins return the oxygenated blood from the lungs to the left atrium. Branches of the pulmonary vessels accompany the bronchi to supply a segment of the lung.

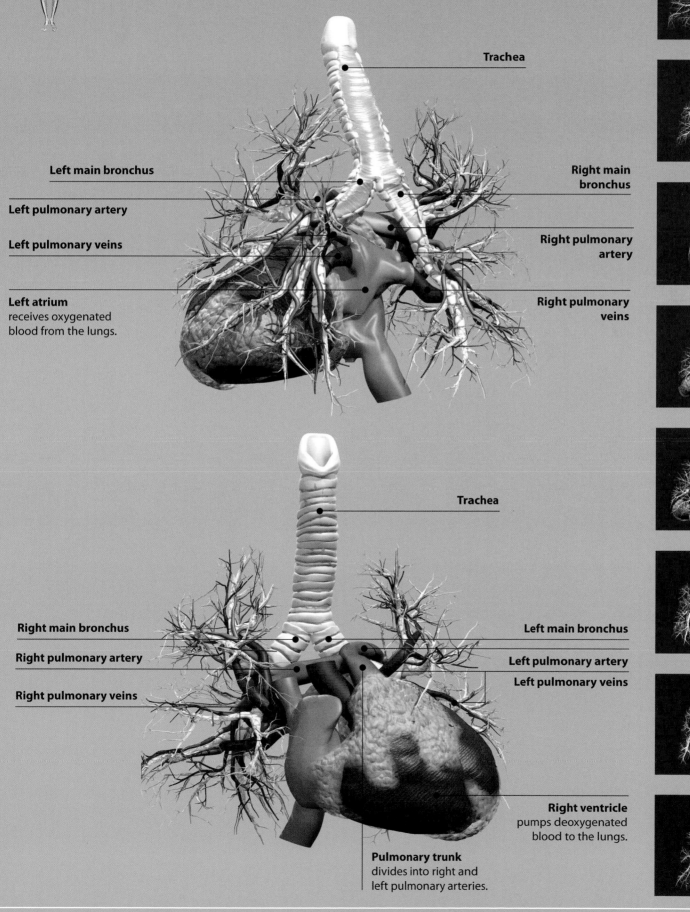

Trachea

Left main bronchus

Left pulmonary artery

Left pulmonary veins

Left atrium
receives oxygenated blood from the lungs.

Right main bronchus

Right pulmonary artery

Right pulmonary veins

Trachea

Right main bronchus

Right pulmonary artery

Right pulmonary veins

Left main bronchus

Left pulmonary artery

Left pulmonary veins

Right ventricle
pumps deoxygenated blood to the lungs.

Pulmonary trunk
divides into right and left pulmonary arteries.

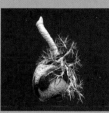

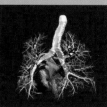

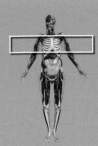

The breasts form paired, variably sized mounds, on the front of the thorax. They are made up of fatty, glandular, and connective tissues. Female hormones such as estrogen cause the breast tissues to grow and develop, particularly at puberty and during pregnancy. The mammary glands of the breast produce milk.

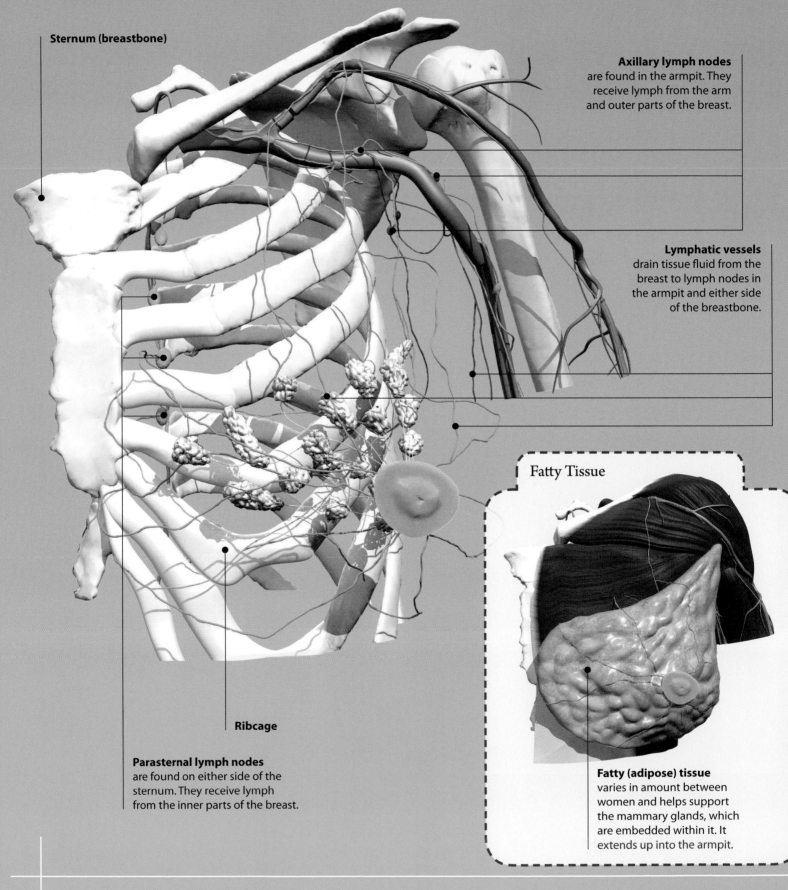

Sternum (breastbone)

Axillary lymph nodes
are found in the armpit. They receive lymph from the arm and outer parts of the breast.

Lymphatic vessels
drain tissue fluid from the breast to lymph nodes in the armpit and either side of the breastbone.

Fatty Tissue

Ribcage

Parasternal lymph nodes
are found on either side of the sternum. They receive lymph from the inner parts of the breast.

Fatty (adipose) tissue
varies in amount between women and helps support the mammary glands, which are embedded within it. It extends up into the armpit.

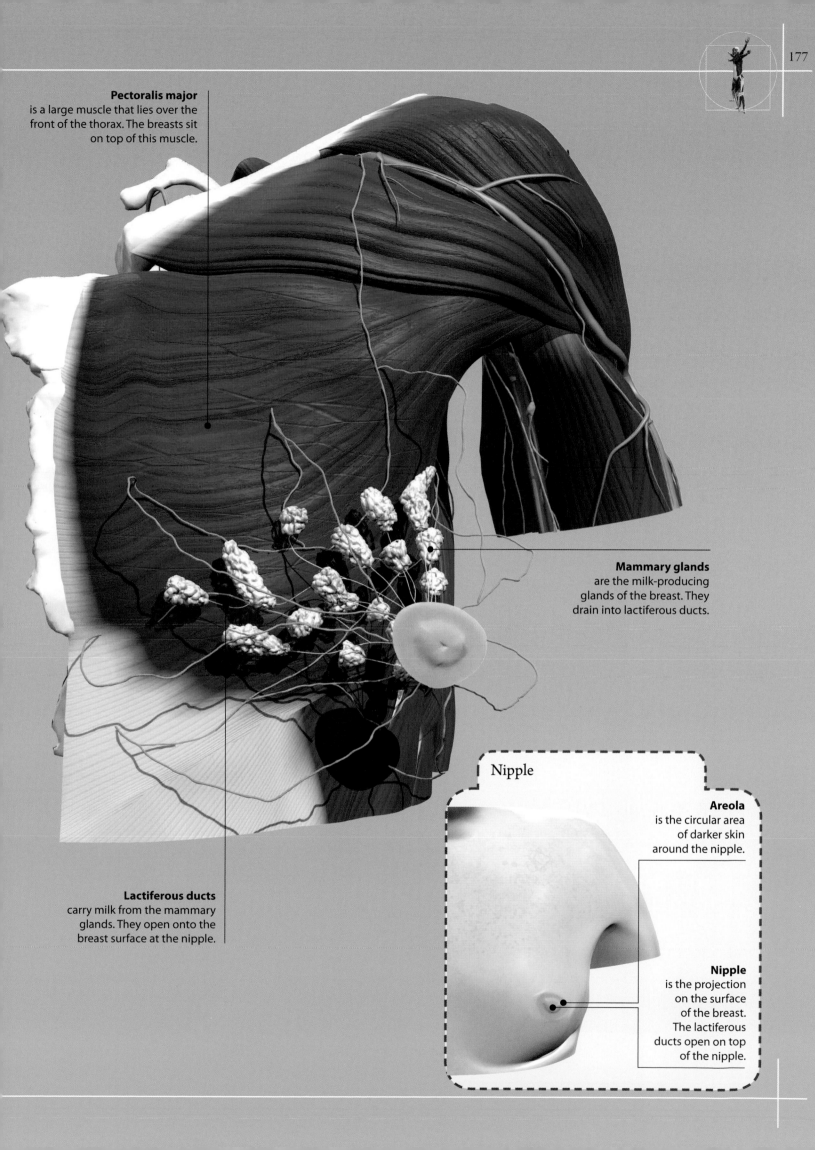

Pectoralis major
is a large muscle that lies over the front of the thorax. The breasts sit on top of this muscle.

Mammary glands
are the milk-producing glands of the breast. They drain into lactiferous ducts.

Lactiferous ducts
carry milk from the mammary glands. They open onto the breast surface at the nipple.

Nipple

Areola
is the circular area of darker skin around the nipple.

Nipple
is the projection on the surface of the breast. The lactiferous ducts open on top of the nipple.

The abdomen forms the lower part of the trunk. It extends from the diaphragm above to the pelvis below. The space inside the abdomen is called the abdominal cavity. It contains most of the organs of the digestive system, including the stomach, liver, pancreas, the small intestine, and part of the large intestine. The kidneys and ureters are found on the back wall of the abdomen.

The walls of the abdomen are mainly formed from layers of muscles, which can bend the trunk forward and from side to side. These muscles are also able to generate large changes in the size and pressure of the abdominal cavity. This helps with bodily functions such as breathing and going to the toilet.

The surface of the abdomen can be divided by vertical and horizontal lines into anatomical regions or quadrants. These areas are associated with different underlying abdominal organs. This information can be used by doctors to accurately describe the position of abdominal pain, and also suggest possible diagnoses.

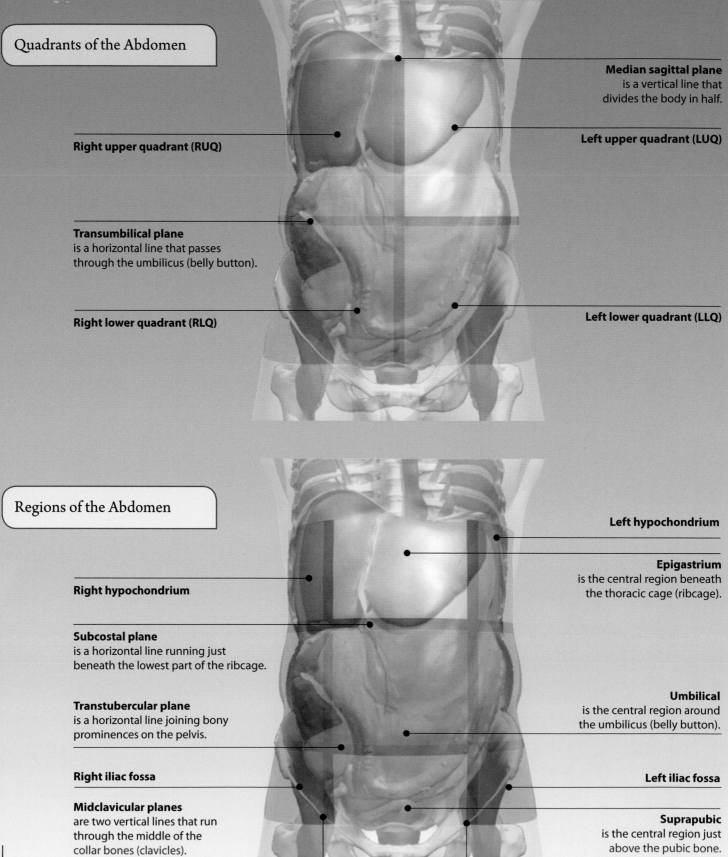

Quadrants of the Abdomen

Median sagittal plane is a vertical line that divides the body in half.

Right upper quadrant (RUQ)

Left upper quadrant (LUQ)

Transumbilical plane is a horizontal line that passes through the umbilicus (belly button).

Right lower quadrant (RLQ)

Left lower quadrant (LLQ)

Regions of the Abdomen

Left hypochondrium

Epigastrium is the central region beneath the thoracic cage (ribcage).

Right hypochondrium

Subcostal plane is a horizontal line running just beneath the lowest part of the ribcage.

Umbilical is the central region around the umbilicus (belly button).

Transtubercular plane is a horizontal line joining bony prominences on the pelvis.

Right iliac fossa

Left iliac fossa

Midclavicular planes are two vertical lines that run through the middle of the collar bones (clavicles).

Suprapubic is the central region just above the pubic bone.

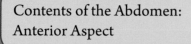

Contents of the Abdomen: Anterior Aspect

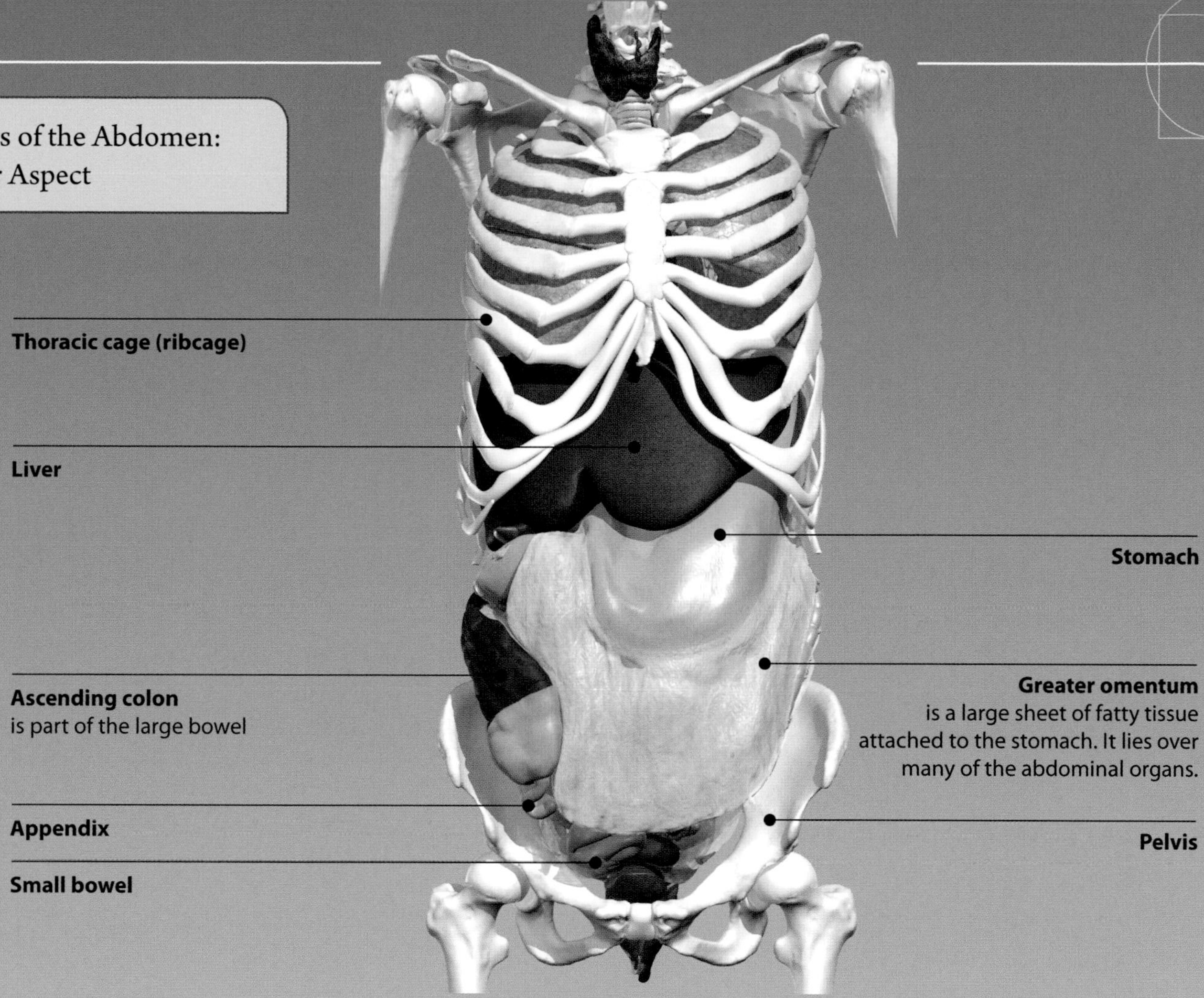

Thoracic cage (ribcage)

Liver

Stomach

Ascending colon
is part of the large bowel

Greater omentum
is a large sheet of fatty tissue
attached to the stomach. It lies over
many of the abdominal organs.

Appendix

Pelvis

Small bowel

Contents of the Abomen: Posterior Aspect

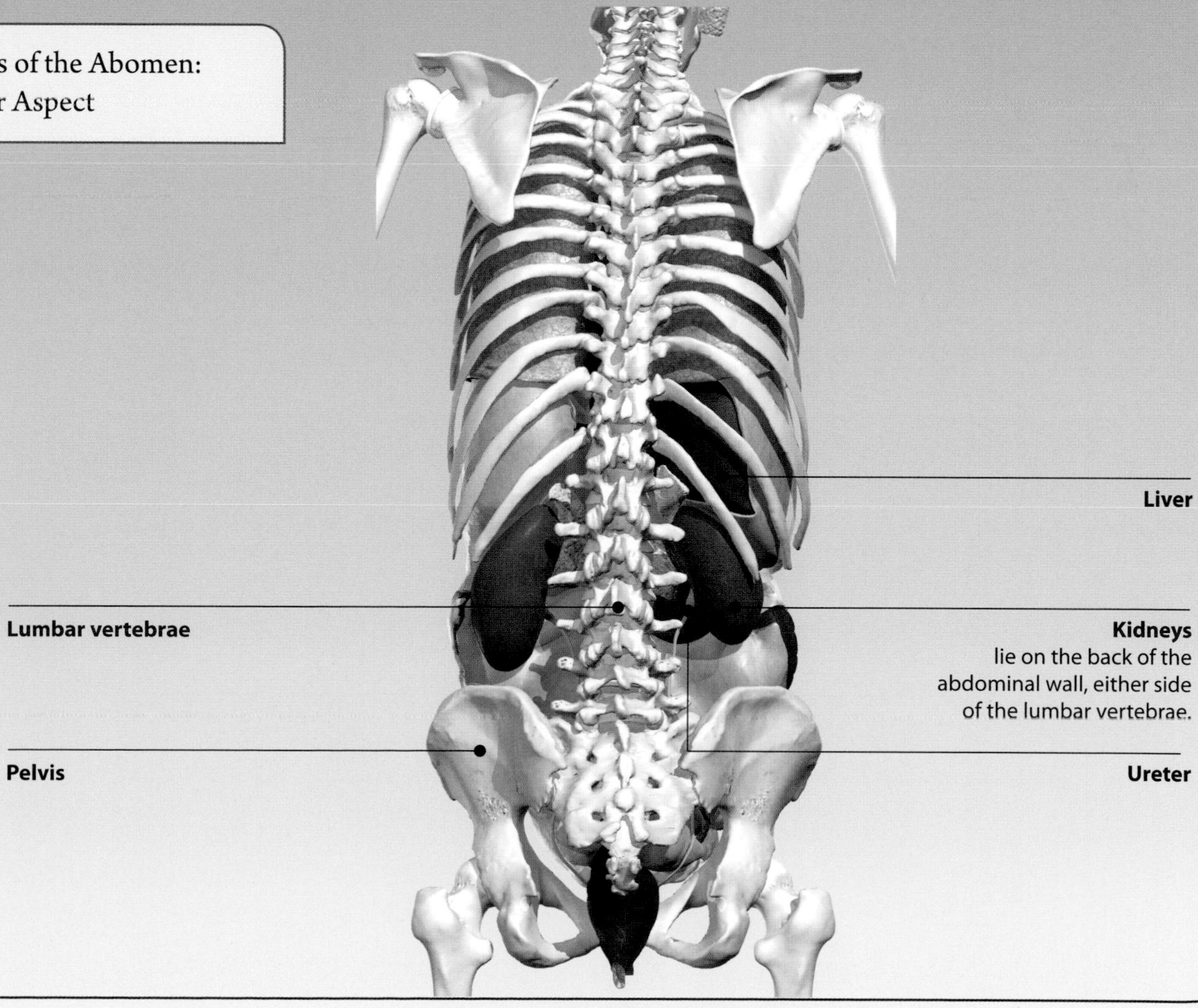

Liver

Lumbar vertebrae

Kidneys
lie on the back of the
abdominal wall, either side
of the lumbar vertebrae.

Pelvis

Ureter

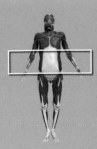

Muscles, and tendon sheets called aponeuroses, form a large part of the abdominal wall. They move the trunk, protect the abdominal contents, and have an important role in increasing the pressure in the abdominal cavity. This is vital for processes such as breathing, childbirth, vomiting, and going to the toilet.

Three flattened muscles—external oblique, internal oblique, and transverse abdominis—form the side walls of the abdomen. Their aponeuroses combine together at the front to enclose the rectus abdominis muscle.

Rectus abdominis
runs from the lower part of the ribcage to the pelvis, forming the front part of the abdominal wall. It bends the trunk forward, allowing us to do "sit-ups."

Tendinous intersections
divide the rectus abdominis muscle at regular intervals. These are what give the "six-pack" appearance.

Transversus abdominis
is the deepest of the three muscles that form the sides of the abdominal wall. It squeezes the abdominal contents when it contracts.

Aponeurosis
is a muscle tendon in the form of a flat sheet.

Linea alba
is a thick band of fibrous tissue that runs the length of the rectus abdominis muscle. It divides it into two halves, and provides attachment for the abdominal wall muscle aponeuroses.

Posterior Abdominal Wall

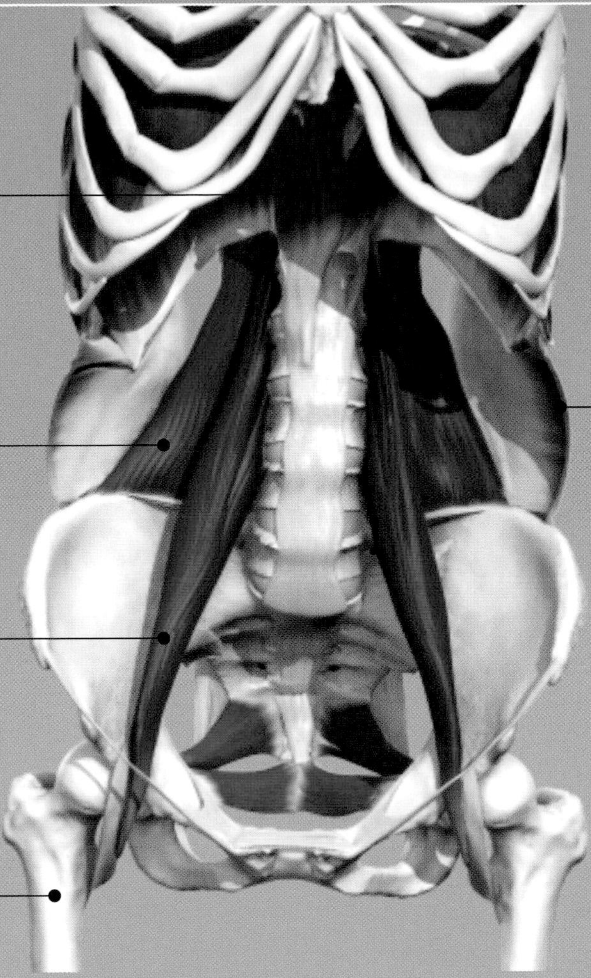

Diaphragm

Quadratus lumborum
forms part of the abdominal
wall at the back. It bends
the trunk to the sides.

Transversus abdominis

Psoas major
is attached to the femur. It
helps bend the trunk forward.

Femur (thigh bone)

Vessels of the Abdominal Wall

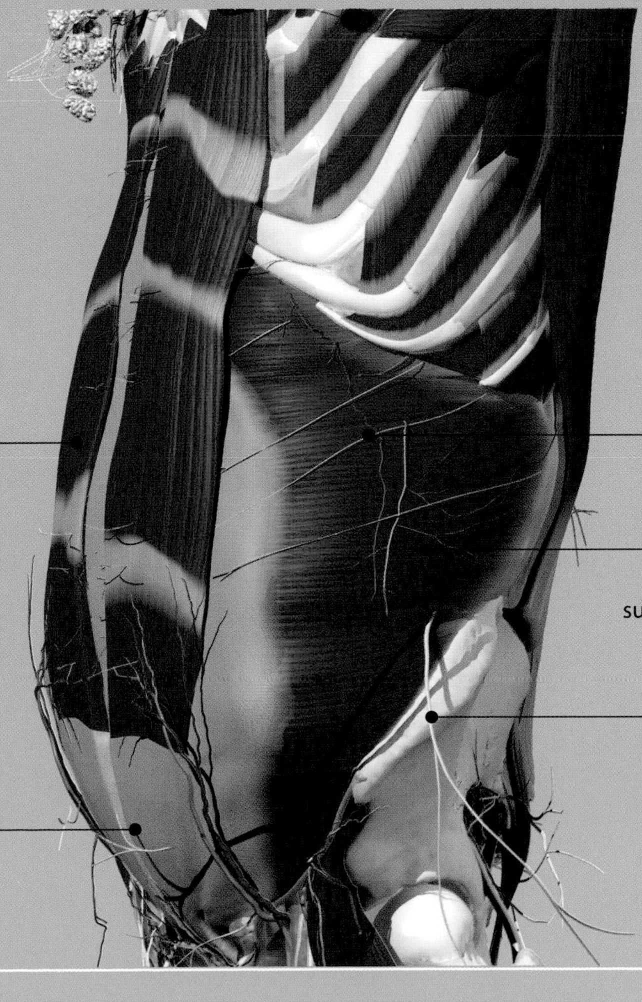

Rectus abdominis

Intercostal nerves
innervate many of the
abdominal wall muscles.

Transversus abdominis

Iliohypogastric nerve
supplies skin over the upper
part of the buttocks and
innervates some of the
abdominal wall muscles.

Superficial epigastric vessels
help supply blood to the
front of the abdominal wall.

The abdominal organs are supplied with blood from branches of the abdominal aorta. The hepatic portal system allows nutrient rich venous blood from the digestive system to be processed by the liver, before draining into the inferior vena cava.

The lumbar plexus is a collection of nerves found at the back of the abdomen. These supply some of the skin of the thigh, leg, and genital region, as well as innervating some of the muscles of the thigh and abdomen.

Vessels of the Abdomen

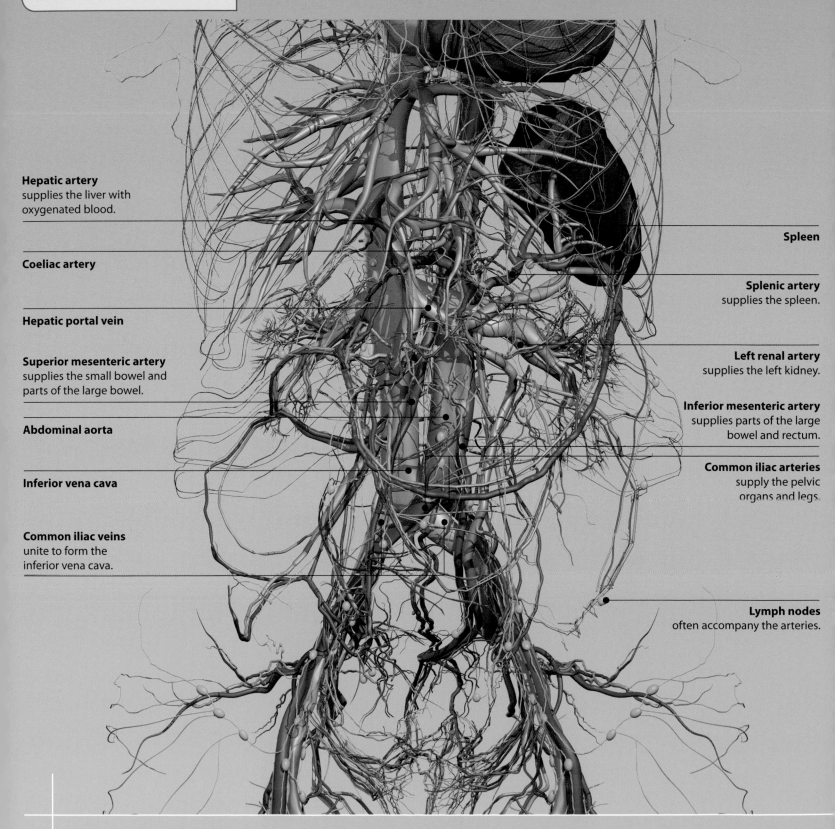

Hepatic artery
supplies the liver with oxygenated blood.

Coeliac artery

Hepatic portal vein

Superior mesenteric artery
supplies the small bowel and parts of the large bowel.

Abdominal aorta

Inferior vena cava

Common iliac veins
unite to form the inferior vena cava.

Spleen

Splenic artery
supplies the spleen.

Left renal artery
supplies the left kidney.

Inferior mesenteric artery
supplies parts of the large bowel and rectum.

Common iliac arteries
supply the pelvic organs and legs.

Lymph nodes
often accompany the arteries.

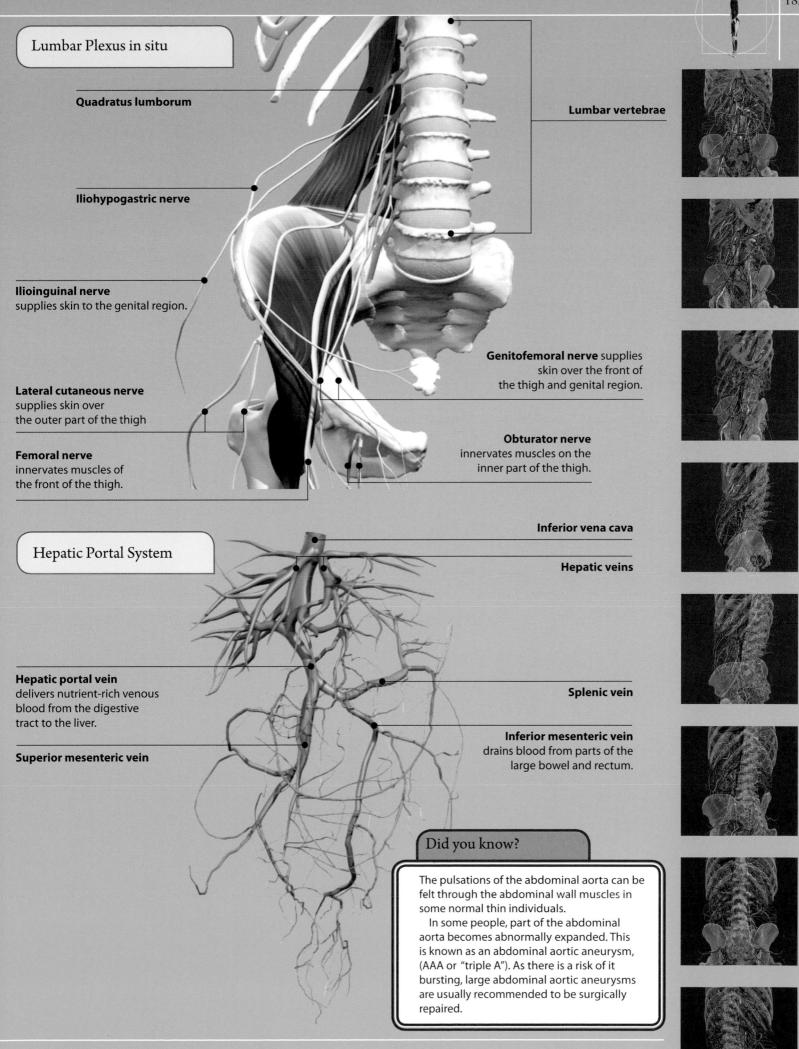

Lumbar Plexus in situ

Quadratus lumborum

Iliohypogastric nerve

Ilioinguinal nerve
supplies skin to the genital region.

Lateral cutaneous nerve
supplies skin over
the outer part of the thigh

Femoral nerve
innervates muscles of
the front of the thigh.

Lumbar vertebrae

Genitofemoral nerve supplies
skin over the front of
the thigh and genital region.

Obturator nerve
innervates muscles on the
inner part of the thigh.

Inferior vena cava

Hepatic Portal System

Hepatic veins

Hepatic portal vein
delivers nutrient-rich venous
blood from the digestive
tract to the liver.

Splenic vein

Inferior mesenteric vein
drains blood from parts of the
large bowel and rectum.

Superior mesenteric vein

Did you know?

The pulsations of the abdominal aorta can be
felt through the abdominal wall muscles in
some normal thin individuals.
  In some people, part of the abdominal
aorta becomes abnormally expanded. This
is known as an abdominal aortic aneurysm,
(AAA or "triple A"). As there is a risk of it
bursting, large abdominal aortic aneurysms
are usually recommended to be surgically
repaired.

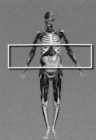

Viewing the abdomen in different anatomical planes allows us to see how the various organs and structure are arranged and related to one another.

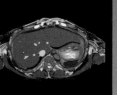

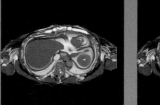

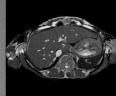

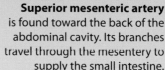

Diaphragm

Pancreas
is closely related to blood vessels draining into the liver.

Stomach
is located near the front of the abdominal cavity.

Greater omentum
loops down from the stomach and transverse colon, forming an apronlike covering for most of the abdominal contents.

Rectus abdominis
is the "six-pack" muscle that makes up the central part of the abdominal wall.

Urinary bladder
is a muscular store for urine.

Hepatic portal vein
delivers nutrient-rich blood from the intestines to the liver.

Superior mesenteric artery
is found toward the back of the abdominal cavity. Its branches travel through the mesentery to supply the small intestine.

Transverse colon
is part of the large intestine.

Rectum

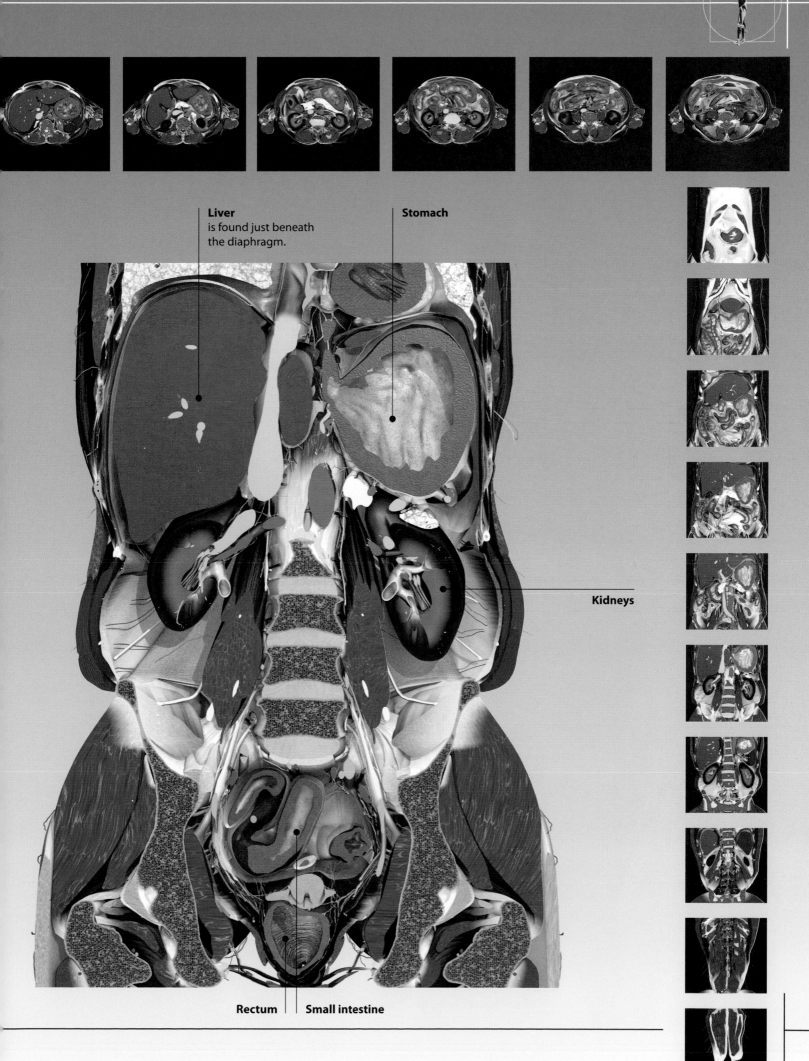

Liver
is found just beneath
the diaphragm.

Stomach

Kidneys

Rectum | **Small intestine**

The peritoneum is a thin, smooth membrane that coats the walls of the abdominal cavity (parietal peritoneum), as well as some of the organs (visceral peritoneum). The peritoneal cavity is the space between the visceral and parietal peritoneum. It contains a small amount of fluid, which helps fight infection and lubricates the movements of the abdominal organs. In places the peritoneum forms layered folds, which attach organs to the abdominal wall and carry their blood vessels.

Liver
is nearly entirely covered by visceral peritoneum.

Falciform ligament
is a two-layered fold of peritoneum that attaches the liver to the abdominal wall.

Ascending colon
is fixed to the abdominal wall by peritoneum that covers its front and side surfaces.

Visceral peritoneum
covers the surfaces of the abdominal organs.

Stomach
is completely covered by visceral peritoneum.

Greater omentum
is a multilayered fold of peritoneum and fatty tissue. It is attached to the stomach and transverse colon, and lies over most of the abdominal contents.

Posterior Abdominal Wall

Parietal peritoneum
lines the walls of the abdominal cavity.

Gallbladder

Kidneys
lie behind the peritoneum.

Diaphragm
has its lower surface covered with peritoneum.

Pancreas
lies behind the peritoneum, and is termed a retroperitoneal structure.

Duodenum
lies mostly behind the peritoneum.

Transverse Mesocolon

Transverse mesocolon
is a double layer of peritoneum that attaches the transverse colon to the abdominal wall at the back.

Ascending colon

Transverse colon
is suspended within the peritoneal cavity by the mesocolon.

Hepatic veins

Pancreas

Blood vessels
that supply the transverse colon run between the layers of the mesocolon.

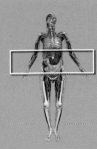

The stomach is a J-shaped organ that connects the esophagus to the duodenum (the first part of the small intestine). It lies on the left side of the upper abdomen, just beneath the diaphragm. It is an expandable, muscular sac, which can store, mix, and break down food to produce a thick liquid called chyme. The delivery of chyme to the duodenum is controlled by the pyloric sphincter.

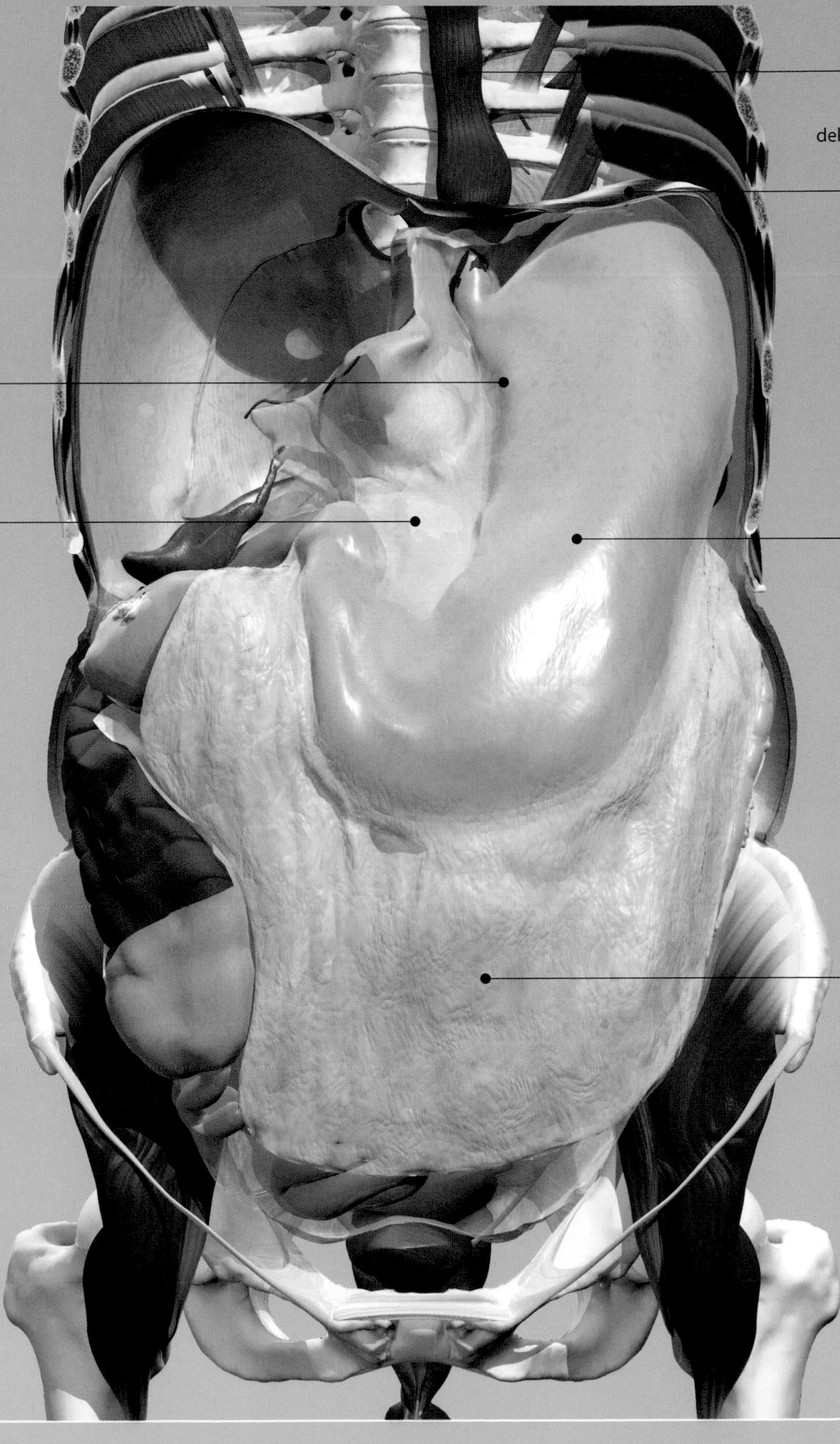

Esophagus
is a muscular tube that delivers food to the stomach.

Diaphragm

Lesser curvature
is the inner curved surface of the stomach.

Lesser omentum
is a fold of peritoneum that connects the stomach to the liver.

Stomach

Greater omentum
is attached to the greater curvature of the stomach.

Internal Surface of Stomach

Cardia
is the region of the stomach
where the esophagus enters.

Fundus
is the upper dome-shaped
region of the stomach.

Body
is the main part
of the stomach.

Pyloric sphincter
is a ring of muscle that varies
the size of the opening from the
stomach into the duodenum.

Pylorus
is the channel that directs chyme
towards the duodenum.

Rugae
are folds in the stomach. They
allow the stomach to expand
and store food after a meal.

Stomach wall
has three muscle layers
(longitudinal, circular, and oblique)
that help mix the food. The lining
cells produce acid and enzymes that
help start digesting food.

Vessels of the Stomach

Esophagus

Aorta

Inferior vena cava

Celiac trunk
is a large blood vessel
arising from the aorta.
Its branches supply the
stomach, liver, and spleen
with arterial blood.

Hepatic portal vein
drains venous blood from the
stomach, via the liver, to the
inferior vena cava.

Gastric arteries
supply the lesser curvature
of the stomach.

Gastroepiploic arteries
supply the greater
curvature of the stomach.

The stomach wall contains specialized cells and tissues that help it to mix and digest food. It is made up of four layers: mucosa, submucosa, muscularis, and serosa.

Stomach Wall

Gastric pits

Gastric glands
contain cells that produce mucus, acid, and enzymes.

Mucosa
is the innermost layer and has three parts: epithelium, lamina propria, and muscularis mucosa.

Epithelium
is the layer of cells lining the stomach. It forms deep folds called gastric pits.

Lamina propria
is a layer of connective tissue that supports the epithelium.

Muscularis
has three muscle layers that help mix the food.

Serosa
is the outermost layer of the stomach, formed by the visceral peritoneum.

Longitudinal muscle fibers
run along the length of the stomach.

Circular muscle fibers
run around the stomach.

Oblique muscle fibers
run diagonally.

Submucosa
is a connective tissue layer that supports the mucosa.

Muscularis mucosa
is a thin layer of muscle that squeezes the gastric glands and pushes their contents onto the stomach surface.

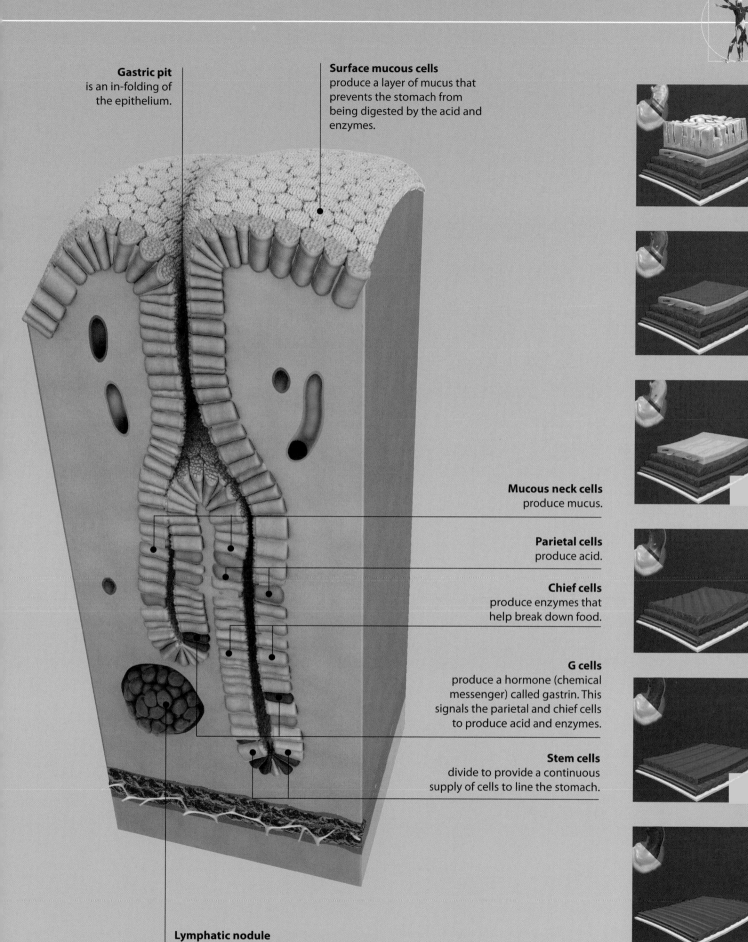

Gastric pit
is an in-folding of
the epithelium.

Surface mucous cells
produce a layer of mucus that
prevents the stomach from
being digested by the acid and
enzymes.

Mucous neck cells
produce mucus.

Parietal cells
produce acid.

Chief cells
produce enzymes that
help break down food.

G cells
produce a hormone (chemical
messenger) called gastrin. This
signals the parietal and chief cells
to produce acid and enzymes.

Stem cells
divide to provide a continuous
supply of cells to line the stomach.

Lymphatic nodule
are collections of cells from the
immune system. They help fight
any infection that may try and
enter via the stomach wall.

Gastric Gland

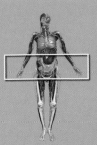

The small intestine is approximately twenty feet long. It is tightly coiled and lies within the abdominal cavity, connecting the stomach and large intestine. It is divided into three sections: duodenum, jejunum, and ileum.

Duodenum
is the first and shortest section of the small intestine. It is connected to the stomach, and forms a C-shape around the head of the pancreas. Bile from the gallbladder and enzymes from the pancreas, are released into the duodenum.

Pancreas
produces digestive enzymes, which are released into the duodenum.

Jejunum
forms the middle section of the small intestine. It is about eight feet long.

Gallbladder
stores and concentrates bile, which is released into the duodenum.

Ileum
is the final section of the small intestine, connected to the large intestine. It is about 12 feet long.

Mesentery
are layers of peritoneum that suspend the small intestine within the abdominal cavity.

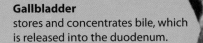

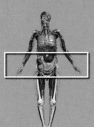

The main roles of the small intestine are digestion (breaking food down into small nutrients) and absorption (taking them into the body). The lining of the small intestine has fingerlike projections called villi, which massively increase the surface area available for absorption.

Microanatomy of the Small Intestine Wall

Villi

Absorptive cells
absorb nutrients from
the small intestine.

Goblet cell
produces mucous that lines and
lubricates the small intestine surface.

Enteroendocrine cells
release hormones which control
secretion of digestive enzymes.

Blood and lymphatic vessels
carry nutrients to and from the small intestine.

Paneth cells
produce substances that
destroy bacteria and protect
the body from infection.

Stem cell
provides a constant supply of cells
to line the small intestine.

Vessels of the Small Intestine

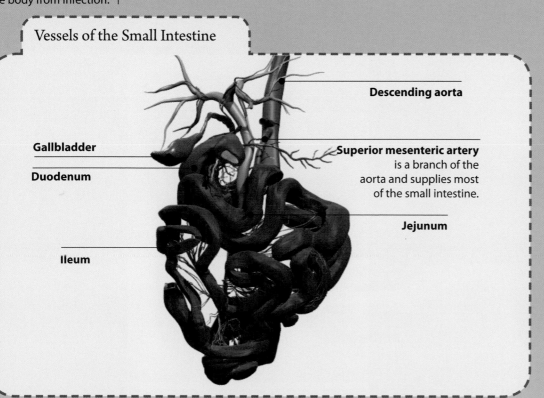

Descending aorta

Gallbladder

Duodenum

Superior mesenteric artery
is a branch of the
aorta and supplies most
of the small intestine.

Jejunum

Ileum

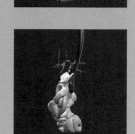

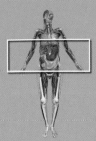

The large intestine is a wide, muscular tube that is about five feet long. It runs from the ileum of the small intestine to the anal canal.

The main functions of the large intestine are the absorption of water and salts from the remaining food, and the formation of waste material. This is removed from the body in a process called defecation.

Did you know?

The normal large intestine contains trillions of bacteria. They help break down any remaining nutrients, and by doing this produce gas (flatus), as well giving feces their distinctive odor and color.

Ascending colon
moves up the right side of the abdominal cavity.

Caecum
is the first part of the large intestine. It receives unabsorbed food material from the ileum.

Appendix
is a wormlike, blind-ending sac connected to the caecum.

Transverse mesocolon
suspends the transverse colon within the abdominal cavity.

Descending colon
moves down the left side of the abdominal cavity towards the pelvis.

Transverse colon
connects the ascending and descending colon.

Microanatomy of the Large Intestine

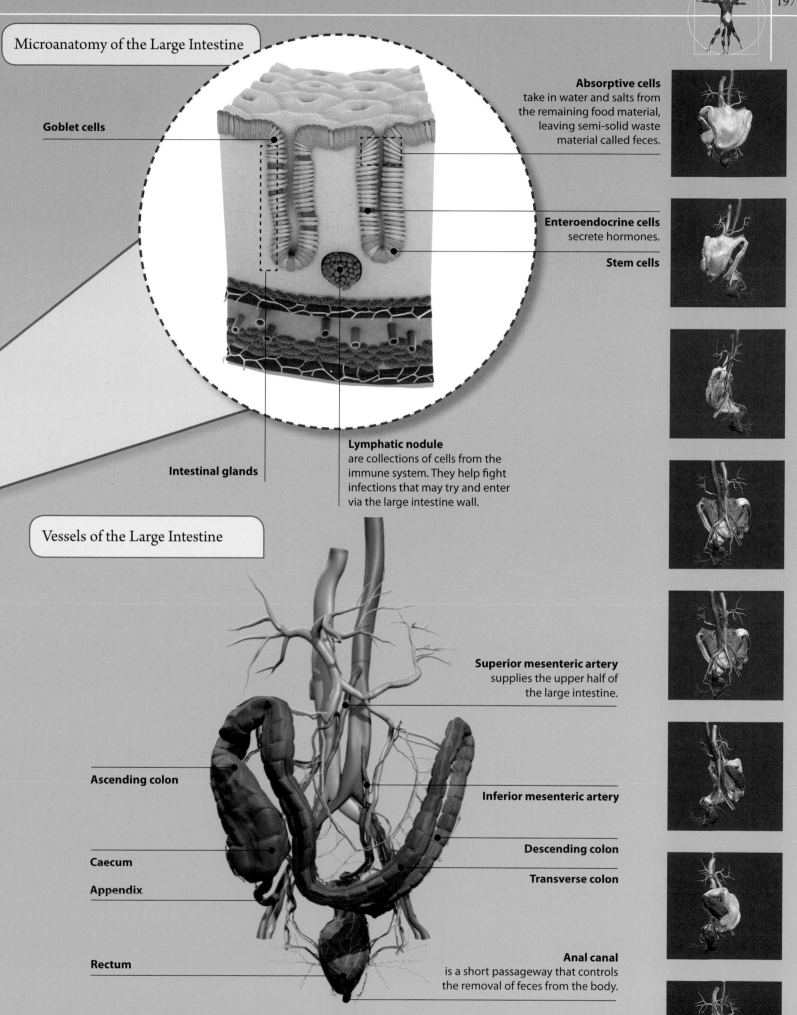

Goblet cells

Absorptive cells
take in water and salts from
the remaining food material,
leaving semi-solid waste
material called feces.

Enteroendocrine cells
secrete hormones.

Stem cells

Intestinal glands

Lymphatic nodule
are collections of cells from the
immune system. They help fight
infections that may try and enter
via the large intestine wall.

Vessels of the Large Intestine

Superior mesenteric artery
supplies the upper half of
the large intestine.

Ascending colon

Inferior mesenteric artery

Caecum

Descending colon

Appendix

Transverse colon

Rectum

Anal canal
is a short passageway that controls
the removal of feces from the body.

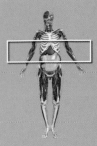

The liver is the largest internal organ in the body. It is located directly beneath the diaphragm, on the right side of the abdomen.

The liver has many important functions. These include processing nutrients absorbed from the small intestine, storage of some vitamins and minerals, producing bile and plasma proteins, and removing poisons from the blood.

The liver has two blood vessels supplying it: the hepatic portal vein and the hepatic artery.

Right lobe
is the largest lobe of the liver.

Left lobe
lies in the middle of the abdomen, covering part of the stomach.

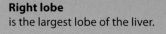

Falciform ligament
attaches the front surface of the liver to the abdominal wall. It is formed from layers of peritoneum, and divides the liver into right and left lobes.

Round ligament
of the liver is a fibrous cord running in the edge of the falciform ligament. It is the remains of the umbilical vein.

Bare Area of the Liver

Caudate lobe
is one of two small lobes
found at the back of the liver.

Right lobe

Left lobe

Bare area
is part of the liver
surface which has no
peritoneum covering it.

Lesser omentum
is a sheet of peritoneum that runs
between the liver and the stomach.
The bile duct, hepatic artery, and
portal vein run within it.

Coronary ligaments
are formed from layers of
peritoneum. They attach the
liver to the diaphragm.

Pancreatic duct
delivers pancreatic enzymes to
the duodenum. It meets with the
common bile duct.

Common bile duct
delivers bile from the gallbladder
to the duodenum. It meets with
the pancreatic duct.

Quadrate lobe
is a small lobe found
below the caudate lobe.

Hepatic Veins

Hepatic veins
drain the liver.

Descending aorta

Hepatic artery
delivers oxygenated
blood to the liver.

Hepatic portal vein
delivers nutrient-rich
blood to the liver for
processing.

Inferior vena cava

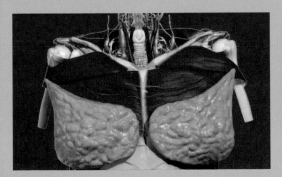

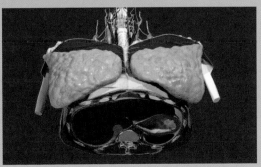

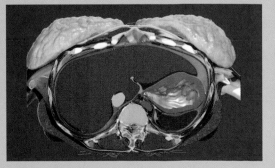

GALLBLADDER AND BILIARY TREE

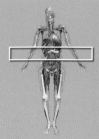

The gallbladder is a small saclike organ found in the right side of the upper abdomen, behind the liver. It stores and concentrates bile. After a fatty meal, the gallbladder squirts bile along the biliary tree and into the small intestine to aid digestion.

Bile is a green liquid produced in the liver. It helps to split up fats and oils so that they can be more easily digested. The biliary tree consists of the ducts, which connect the liver, gallbladder, and pancreas to the duodenum, where bile is released.

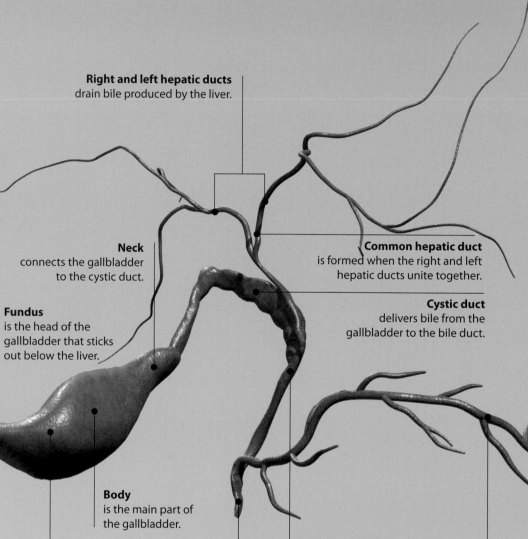

Right and left hepatic ducts
drain bile produced by the liver.

Neck
connects the gallbladder
to the cystic duct.

Common hepatic duct
is formed when the right and left
hepatic ducts unite together.

Fundus
is the head of the
gallbladder that sticks
out below the liver.

Cystic duct
delivers bile from the
gallbladder to the bile duct.

Body
is the main part of
the gallbladder.

Gallbladder
has a fundus, body, and neck.

Common bile duct
is formed when the cystic
and common hepatic ducts
join together. It travels in
the lesser omentum to
meet the pancreatic duct.

Pancreatic duct
carries pancreatic enzymes from
the pancreas to the duodenum.

Hepatopancreatic ampulla
is a short duct formed by
the common bile duct and
pancreatic duct joining together.
It opens into the duodenum,
releasing bile and pancreatic
enzymes to help digestion.

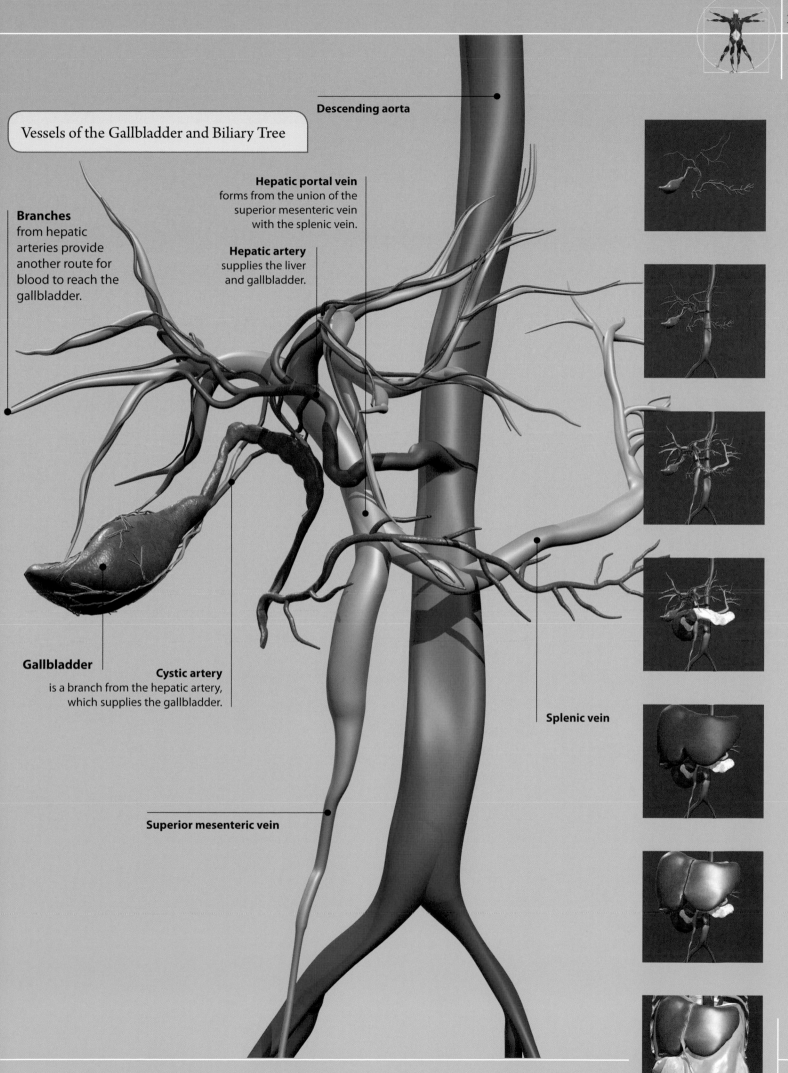

Descending aorta

Vessels of the Gallbladder and Biliary Tree

Hepatic portal vein
forms from the union of the
superior mesenteric vein
with the splenic vein.

Branches
from hepatic
arteries provide
another route for
blood to reach the
gallbladder.

Hepatic artery
supplies the liver
and gallbladder.

Gallbladder

Cystic artery
is a branch from the hepatic artery,
which supplies the gallbladder.

Splenic vein

Superior mesenteric vein

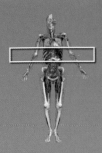

The pancreas is a long, thin organ that lies horizontally along the back wall of the upper abdomen behind the stomach. It produces enzyme-rich pancreatic juice, which is released into the duodenum to help digestion. The pancreas also produces the hormones insulin and glucagon, which help control blood sugar levels.

The spleen is an oval-shaped organ, and is part of the lymphatic system. It lies on the upper left side of the abdomen, between the stomach and the diaphragm. It contains large numbers of white blood cells and filters the blood.

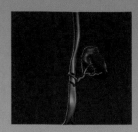

Spleen

Descending aorta

Spleen
is the largest single lymphoid organ in the body. It filters the blood, removing invading organisms and worn-out blood cells.

Celiac trunk
is one of the large branches of the aorta within the abdomen.

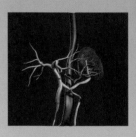

Hepatic portal vein

Splenic vein
drains blood from the spleen. It unites with the superior mesenteric vein to form the hepatic portal vein.

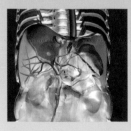

Splenic artery
is the branch of the celiac trunk, which supplies the spleen.

Superior mesenteric vein

Inferior mesenteric vein
drains into the splenic vein.

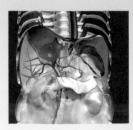

Pancreas, Duodenum, and Pancreatic Duct

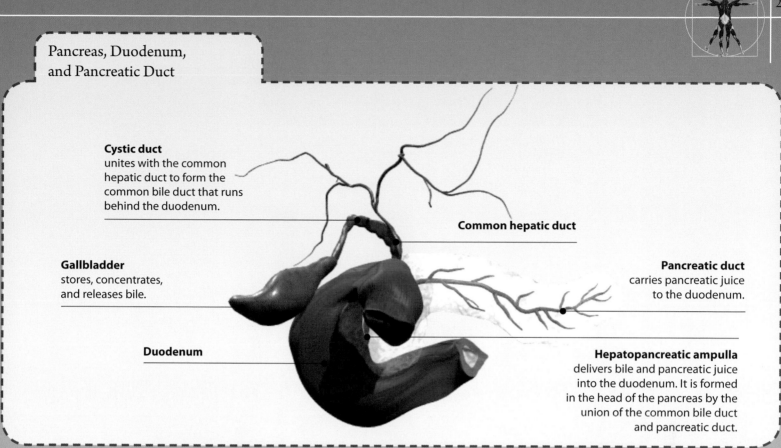

Cystic duct
unites with the common hepatic duct to form the common bile duct that runs behind the duodenum.

Common hepatic duct

Gallbladder
stores, concentrates, and releases bile.

Pancreatic duct
carries pancreatic juice to the duodenum.

Duodenum

Hepatopancreatic ampulla
delivers bile and pancreatic juice into the duodenum. It is formed in the head of the pancreas by the union of the common bile duct and pancreatic duct.

Pancreas and Spleen within the Abdomen

Gallbladder

Head
of the pancreas lies within the curve of the C-shaped duodenum.

Spleen

Body
of the pancreas forms the middle section of the pancreas.

Tail
of the pancreas is found on the left, next to the spleen and left kidney.

Kidneys

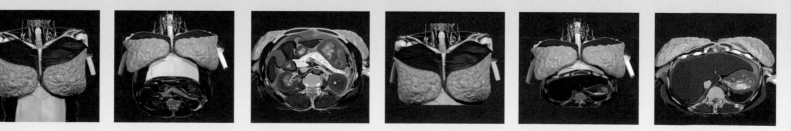

The kidneys are two bean-shaped organs, found on either side of the spine. They lie behind the peritoneum (retroperitoneal), high up on the back wall of the abdomen.

The kidneys filter the blood of impurities, forming a waste product called urine. They control the amount of water and salts in the body, as well as producing hormones that control blood pressure, red blood cell production, and calcium levels.

Aorta

Suprarenal (adrenal) glands
are found above the kidneys. They produce steroid hormones and adrenaline, which help us deal with stressful situations.

Renal veins
drain blood from the kidneys.

Renal artery
are branches from the aorta that bring blood to the kidneys to be filtered.

Cortex
is the outer part of the kidney where blood is filtered.

Medulla
is the inner part of the kidney that concentrates the urine.

Calyces
are horn-shaped tubes that collect the urine produced by each pyramid.

Pyramids
are found in the medulla, and drain urine into the calyces.

Ureters
are muscular tubes that carry urine from the kidneys to the bladder.

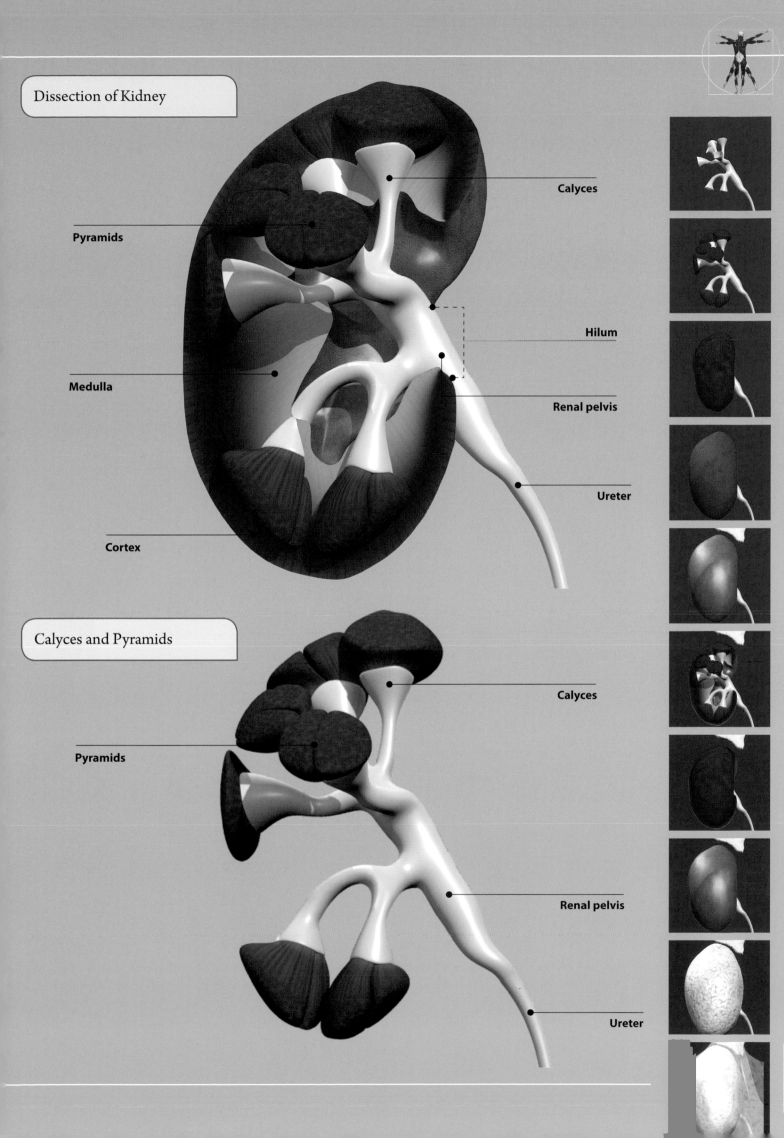

Dissection of Kidney

Calyces

Pyramids

Hilum

Medulla

Renal pelvis

Ureter

Cortex

Calyces and Pyramids

Calyces

Pyramids

Renal pelvis

Ureter

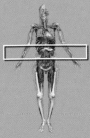

Each kidney contains about a million nephrons. These are the working units of the kidney, which filter the blood and process the fluid, eventually forming urine. This drains into the calyces at the tip of the pyramids. Each nephron consists of a renal corpuscle, a renal tubule, and a collecting duct.

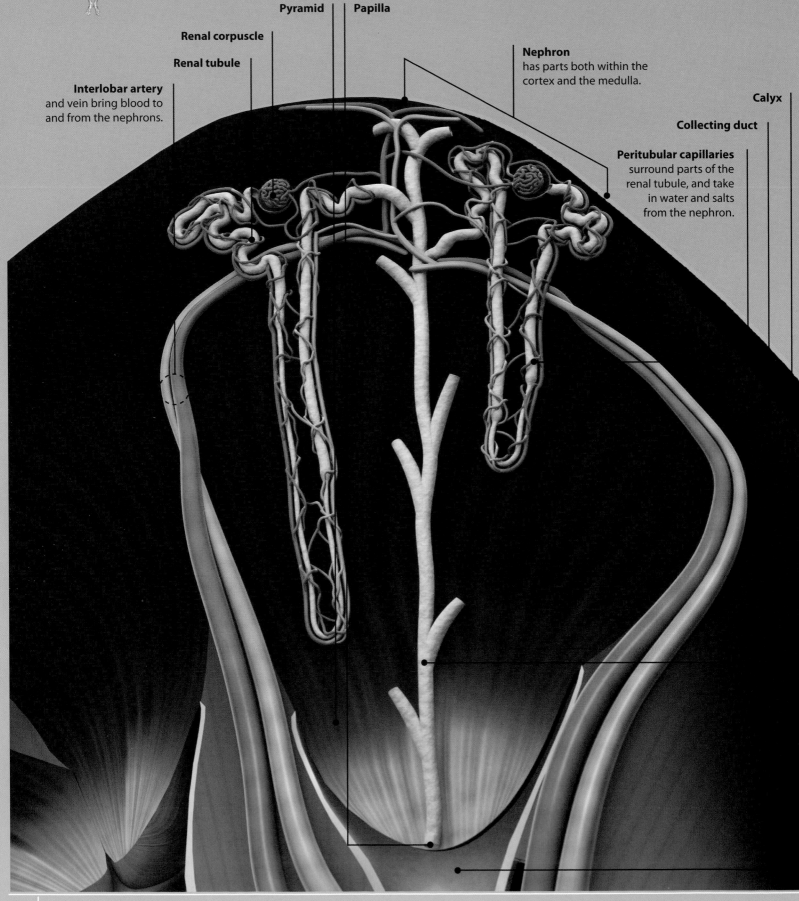

Pyramid

Papilla

Renal corpuscle

Renal tubule

Interlobar artery
and vein bring blood to
and from the nephrons.

Nephron
has parts both within the
cortex and the medulla.

Calyx

Collecting duct

Peritubular capillaries
surround parts of the
renal tubule, and take
in water and salts
from the nephron.

Cells of the Nephron

Renal corpuscle
is where the blood is filtered, to form a fluid called the filtrate. It consists of a tight network of capillaries (glomerulus), and a surrounding capsule of cells.

Proximal convoluted tubule
is the twisted first part of the renal tubule. It reabsorbs much of the water and salts from the filtrate.

Loop of Henle
is the second part of the tubule. It dips down into the medulla, before doing a U-turn, and rising up again towards the cortex. This arrangement helps concentrate the urine.

Thin descending limb of the loop of Henle
mainly allows water to pass through its walls.

Distal convoluted tubule
is the twisted third part of the tubule that reabsorbs water and salts.

Collecting duct
receives urine from a number of nephrons and delivers it to the renal papilla, where it empties into the calyces.

Renal Pyramids

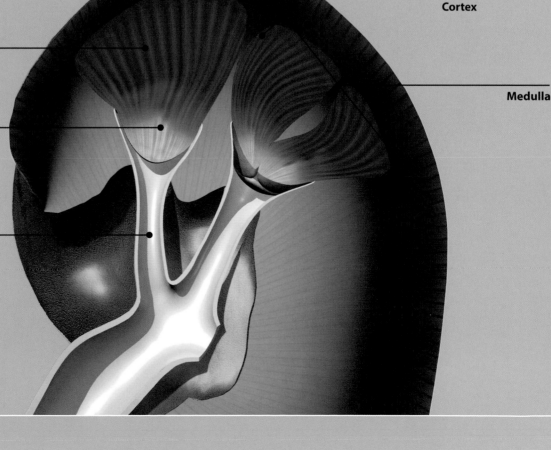

Pyramids
are packed with parts of the tubules and collecting ducts of numerous different nephrons.

Papillae
are the tips of the pyramids that project into the calyces.

Calyces
collect the urine produced by the nephrons in each pyramid.

Cortex

Medulla

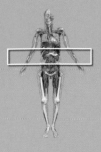

The renal corpuscle is the specialized part of the nephron that filters the blood. It has two parts: a tight ball of capillaries called the glomerulus, and a cup-shaped Bowman's capsule, which surrounds the glomerulus.

Specialized cells called podocytes cover the glomerular capillaries. Together they form a specialized filtration barrier, which only allows certain substances to pass through from the blood, and form a filtrate within Bowman's capsule.

Distal convoluted tubule runs next to the renal corpuscle. Cells from it contribute to the juxtaglomerular apparatus.

Efferent arteriole takes blood away from the glomerulus.

Juxtaglomerular apparatus are a collection of specialized cells that help control blood pressure and the rate of filtration at the glomerulus.

Proximal convoluted tubule is the first part of the renal tubule.

Afferent arteriole brings blood to the glomerulus.

Glomerular capillaries

Capsular space is the space between the visceral and parietal layers, where the filtrate forms before draining into the renal tubules.

Glomerulus is a tightly coiled collection of capillaries.

Glomerular Capillary

Podocyte foot
processes extend out to cover
the glomerular capillaries.

Podocyte
are specialized cells that
form the visceral layer of
Bowman's capsule, covering
the glomerular capillaries.

Filtration barrier

Fenestrations
are small holes within the
glomerular capillary walls.

Red blood cells

Filtration Barrier

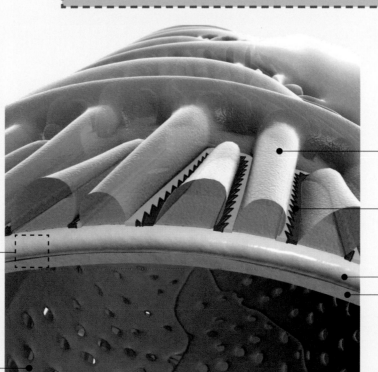

Pedicels
are extensions from the podocyte
foot processes. They interlink
with each other, forming narrow
filtration slits between them.

Filtration barrier
is formed by the podocytes,
basement membrane, and
glomerular capillaries.

Filtration slits

Basement membrane
supports the podocytes and
capillary endothelium.

Glomerular capillary
is one cell thick and forms part of
the filtration barrier.

Fenestrations

The ureters are two muscular tubes, which carry urine from the kidneys to the bladder. They are approximately ten inches long, and twentieth of an inch wide. They travel down the muscles of the posterior abdominal wall, before passing into the pelvis. The kidneys and ureters are held in place on the abdominal wall by the renal fascia.

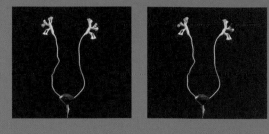

Ureters

Kidneys
produce urine, which drains into the ureters.

Aorta
slits into two common iliac arteries at its lowest point.

Ureters
have layers of muscle that generate "waves," which sweep the urine down to the bladder.

Common iliac arteries
are crossed by the ureters as they divide into internal and external iliac arteries.

Pelvis
contains the bladder. The ureters run along the pelvic side walls before turning in to enter the bladder.

Bladder
collects and stores urine delivered by the ureters.

Urethra
is a tube that expels the urine stored in the bladder from the body.

Psoas major
muscles run either side of the lower spine, down to the thigh bone (femur). The ureters travel on their surface in the upper part of their course.

External iliac artery

Internal iliac artery

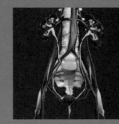

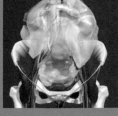

Renal Fascia

Renal fascia
is formed from fibrous tissue.
It supports the kidneys and
ureters in place on the back
abdominal wall.

Kidneys
are held to the back abdominal
wall by the renal fascia.

Ureters

Bladder

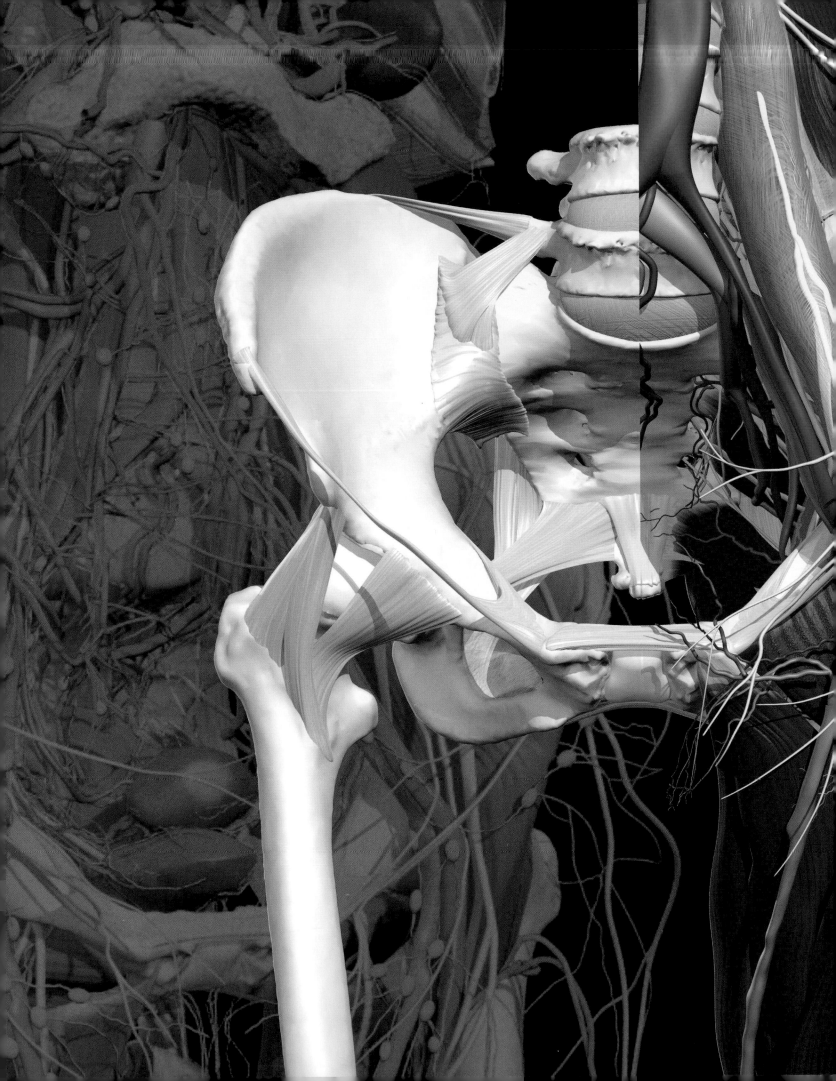

The pelvis is the strong, bowl-shaped area at the base of the trunk, where the legs attach to the rest of the body. It is formed from the hip and lower back bones. It contains and protects the reproductive organs and lower parts of the urinary and digestive systems. A muscular sheet provides support to these organs and forms the pelvic floor.

The pelvis provides attachment for various muscles that move both the trunk and the legs. The shape and position of the pelvis helps us in standing, walking, and keeping an upright position.

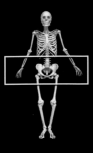

The pelvis is formed by the two hip bones, and the sacrum and coccyx at the back. The hip bone is formed by three separate bones (ilium, ischium, and pubis), which fuse together. They all meet at the acetabulum. The central bowl-shaped region protected by the bones is known as the pelvic cavity, and contains the pelvic organs.

The hip bones form the pelvic girdle, which is the point of attachment of the lower limbs to the rest of the body.

Ligaments
connect the bones of the pelvis, holding them strongly together.

Sacrum

Ilium
has broad bony crests. These form the hips that can be easily felt at the side.

Acetabulum
is a cup-shaped depression in the hip bones. The rounded upper end of the thigh bone fits into this, to form the hip joint.

Pubic symphysis
is the joint that connects the two pubic bones.

Ischium

Pubis (pubic bone)
forms the front of the pelvis.

Femur (thigh bone)
forms the hip joint at its upper end.

Pelvis: Anterior Aspect

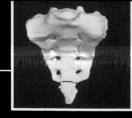
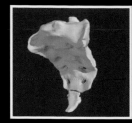

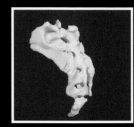
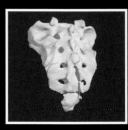

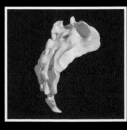

Did you know?

In diseases that affect the blood cells, it is often helpful for doctors to analyze a sample of the bone marrow. A common site for obtaining this sample is either side of the lower back, where the hip bone is near the surface. A special needle is passed into the bone, and a sample of marrow is obtained.

Pelvis: Posterior Aspect

Sacrum
is one of the regions of the backbone (vertebral column). It forms the back wall of the pelvis.

Ilium

Pubic symphysis

Coccyx
is found at the bottom of the sacrum.

Ischium
is the part of the hip bone that we sit on.

Femur (thigh bone)

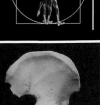

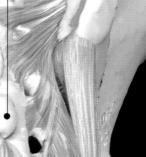

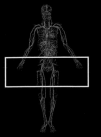

The descending aorta splits into two common iliac arteries just above the pelvis. These vessels and their branches supply the pelvic organs and the lower limbs.

Although most blood vessels are the same in males and females, there are some variations that reflect the differences in the reproductive organs.

In general, the veins of the pelvis follow the same pattern as the arteries.

Arteries of the Female Pelvis

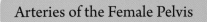

Descending aorta
divides into right and left common iliac arteries.

Ovarian arteries
supply the ovaries directly from the aorta.

Common iliac arteries

Inferior epigastric artery

Superior rectal artery
is a branch from the inferior mesenteric artery that supplies the rectum.

Inferior mesenteric artery continues into the pelvis to supply the rectum.

Internal iliac artery
branches supply the pelvic organs.

External iliac artery
branches supply most of the lower limb and lower abdominal wall.

Vaginal artery
supplies the vagina.

Uterine artery
supplies the uterus (womb).

Internal pudendal artery

Vesical arteries
supply the bladder.

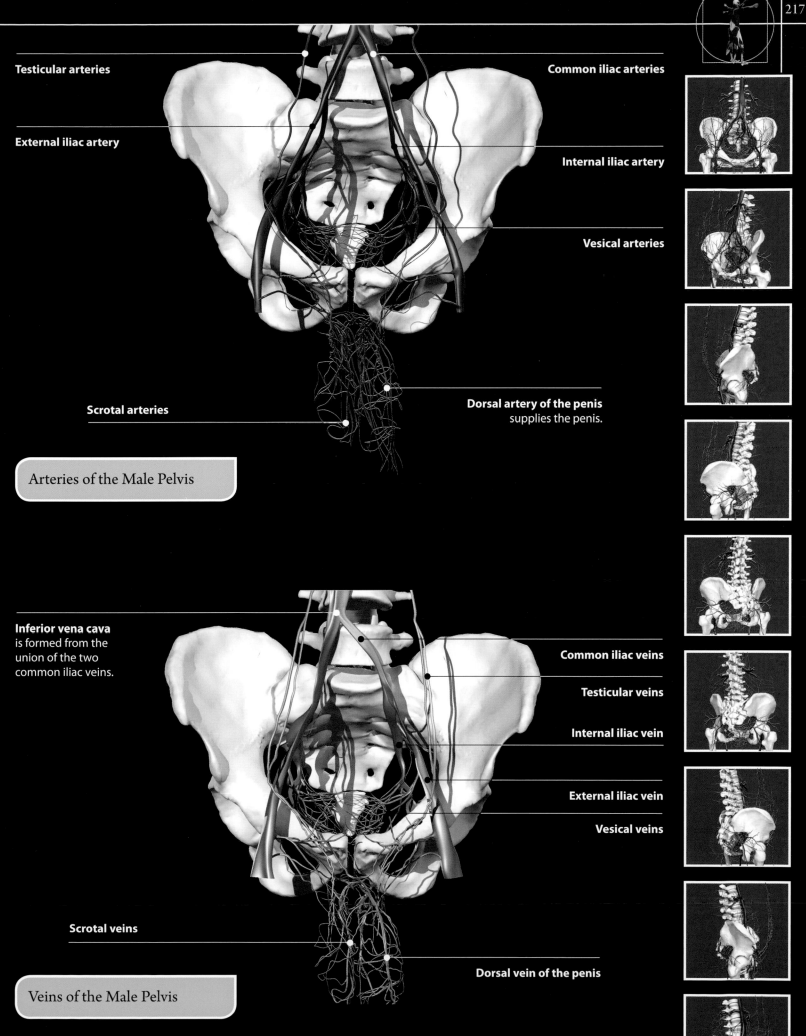

Testicular arteries

External iliac artery

Common iliac arteries

Internal iliac artery

Vesical arteries

Scrotal arteries

Dorsal artery of the penis
supplies the penis.

Arteries of the Male Pelvis

Inferior vena cava
is formed from the
union of the two
common iliac veins.

Common iliac veins

Testicular veins

Internal iliac vein

External iliac vein

Vesical veins

Scrotal veins

Dorsal vein of the penis

Veins of the Male Pelvis

The pelvis contains the nerves that form the sacral plexus, along with branches from the lumbar plexus. Between them, these collections of nerves supply the organs, muscles, and skin of the pelvis and lower limb.

Nerves of the Female Pelvis

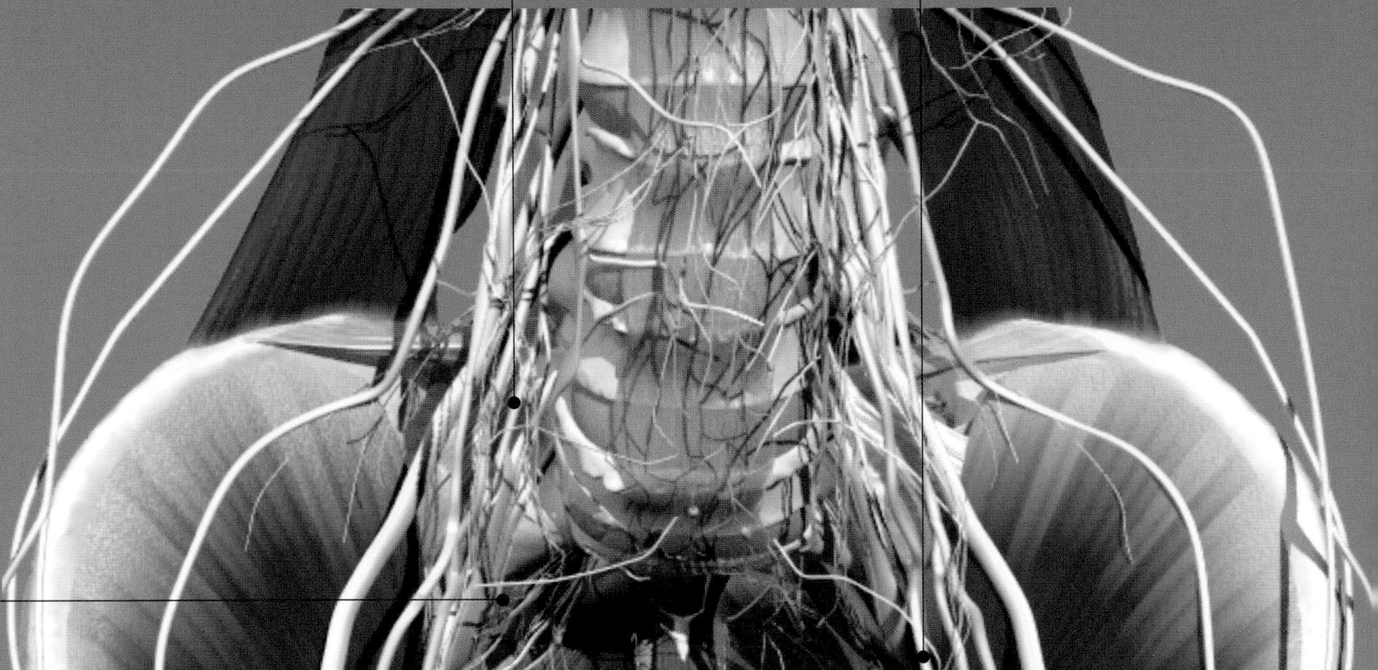

Lumbosacral trunk
is a contribution of nerve fibers from the lumbar plexus to the sacral plexus.

Sciatic nerve

Piriformis

Coccygeus
is a muscle that forms part of the pelvic floor at the back.

Femoral nerve
supplies the muscles at the front of the thigh.

Pelvic floor
is formed from a sheet of muscles. It supports the pelvic organs and provides openings for the urethra, vagina, and anus to reach the outside.

Inferior hypogastric plexus
provides innervation to the pelvic organs.

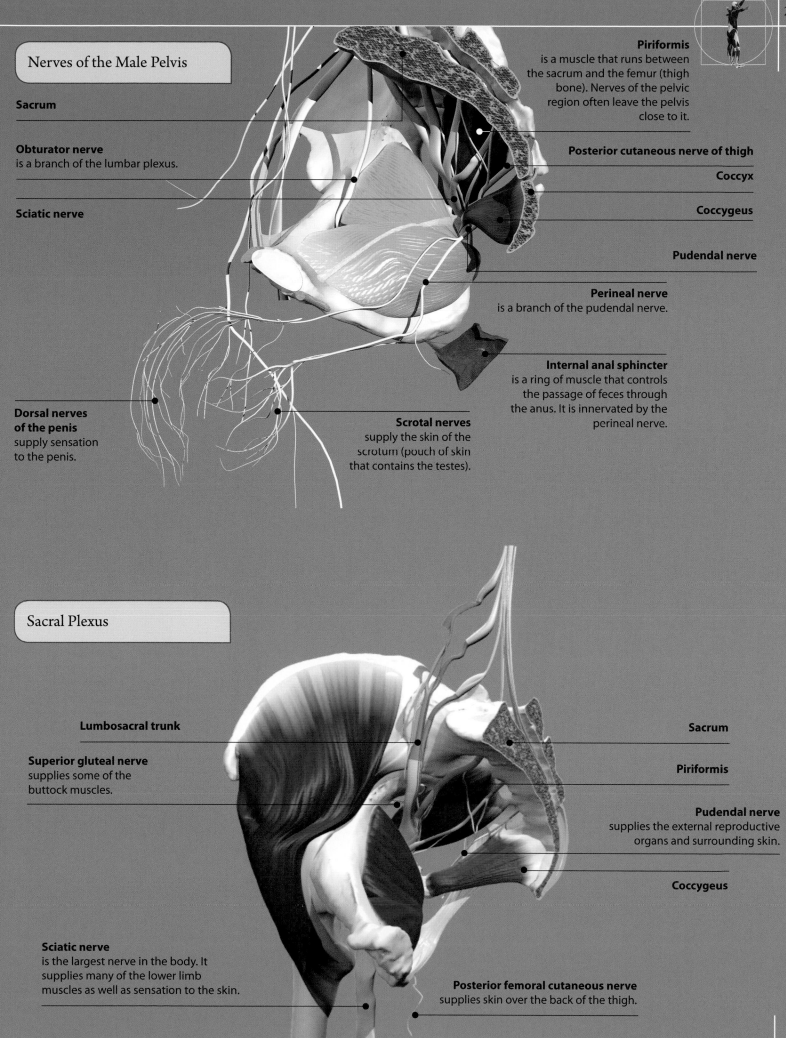

Nerves of the Male Pelvis

Sacrum

Obturator nerve
is a branch of the lumbar plexus.

Sciatic nerve

**Dorsal nerves
of the penis**
supply sensation
to the penis.

Piriformis
is a muscle that runs between
the sacrum and the femur (thigh
bone). Nerves of the pelvic
region often leave the pelvis
close to it.

Posterior cutaneous nerve of thigh

Coccyx

Coccygeus

Pudendal nerve

Perineal nerve
is a branch of the pudendal nerve.

Internal anal sphincter
is a ring of muscle that controls
the passage of feces through
the anus. It is innervated by the
perineal nerve.

Scrotal nerves
supply the skin of the
scrotum (pouch of skin
that contains the testes).

Sacral Plexus

Lumbosacral trunk

Superior gluteal nerve
supplies some of the
buttock muscles.

Sciatic nerve
is the largest nerve in the body. It
supplies many of the lower limb
muscles as well as sensation to the skin.

Sacrum

Piriformis

Pudendal nerve
supplies the external reproductive
organs and surrounding skin.

Coccygeus

Posterior femoral cutaneous nerve
supplies skin over the back of the thigh.

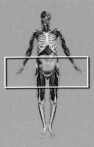

The female reproductive system includes the ovaries, uterine (Fallopian) tubes, uterus, and vagina. Every month from puberty (usually early teens) to menopause (usually early fifties), it undergoes a series of events that prepare it for the possibility of pregnancy. These changes are brought about by hormones, such as estrogen and progesterone.

Every month, a female sex cell (ovum) is released from the ovaries into the uterine tubes. If the ovum is fertilized by a sperm (male sex cell) it settles into the lining of the uterus (womb), where it grows and develops into a baby. After nine months, strong contractions of the uterus push the baby out in the process of childbirth.

Uterine (Fallopian) tube
carries ova to the uterus. This is where fertilization usually occurs.

Ovary
stores and releases female sex cells (ova).

Uterus (womb)
is a hollow muscular organ. Fertilized eggs settle and develop in the lining of the uterus.

Pelvic bone

Labia minora

Cervix
is the lower part of the uterus that opens into the vagina.

Rectum
lies behind the vagina and uterus.

Vagina
is a muscular tube that connects the cervix to the outside world.

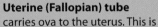

Did you know?

Infection of the uterine tubes is called salpingitis. Following infection, scar tissue may form, which blocks the uterine tubes and prevents the egg and sperm from meeting, resulting in infertility.

Pelvic bones
protects the organs.

Pelvic cavity
is the space inside the pelvis where most of the female reproductive system is found.

Pelvic floor
is a sheet of muscles that support the female reproductive organs.

Internal iliac artery
and its branches supply most of the female reproductive system with blood.

Uterine (Fallopian) tubes

Ovary

Uterus

Urinary bladder
lies in front of the uterus.

Cervix

Vagina

Labia majora are thick folds of skin on either side of the openings of the female reproductive and urinary systems.

Labia minora
are fleshy folds of tissue within the labia majora.

FEMALE PELVIC FLOOR

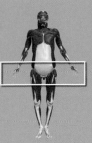

The pelvic floor is a muscular sheet that forms the base of the pelvic cavity. It is made up of numerous muscles, suspended from the pelvic bones. These form a hammocklike support for the pelvic organs. In the female it has openings for the urethra, vagina, and anus.

Female Pelvic Floor:
Superior Aspect

Sacrum & coccyx
form the back wall of the pelvis.

Coccygeus
is a muscle that forms the back of the pelvic floor, running between the coccyx and the hip bone.

Hip bone
provides attachment for many of the pelvic floor muscles.

Opening for anus
Internal anal sphincter muscles surround and compress the upper part of the anal canal.

Levator ani
is a large muscle which forms the majority of the pelvic floor.

Sphincter urethrae muscle
surrounds and compresses the urethra, allowing us to control when to urinate.

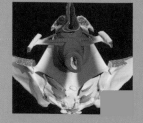
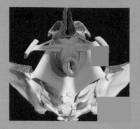
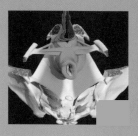

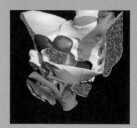
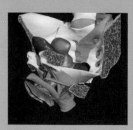
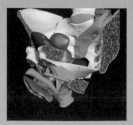

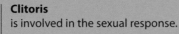

Female Pelvic Floor: Inferior Aspect

Clitoris
is involved in the sexual response.

Bulbospongiosus muscle
passes either side of the vagina,
narrowing its opening.

Ischiocavernosus
covers part of the clitoris.

**Superficial and deep transverse
perineal muscles**
attach the perineal body
to the hip bones.

Hip bone

Ischial tuberosity
is the part of the hip
bone that we sit on.

Levator ani

External anal sphincter muscle
surrounds and compresses the anal
canal, giving us voluntary control
over the process of defecation.

Perineal body
is a tough fibrous structure lying
between the vagina and the anus. It
provides a central point of attachment
for the pelvic floor muscles.

Anal canal   **Vagina**

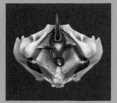
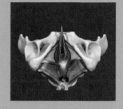

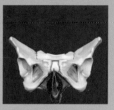
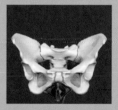
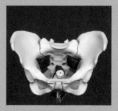
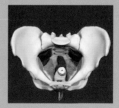
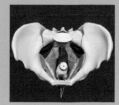
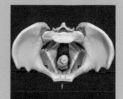

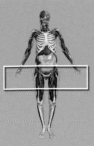

The ovaries are paired almond-shaped structures located within the peritoneal cavity, close to the entrance to the uterine tubes. At birth, the ovaries contain all the developing female gametes, called ova (eggs). These are surrounded by specialized cells to form follicles. Every month during the reproductive years, a few follicles mature and grow until one is large enough to rupture, releasing its ovum in a process called ovulation.

Germinal epithelium
is a thin layer of cells that coats the surface of the ovary.

Cortex
is the outer part of the ovary. It contains the developing follicles.

Primordial follicles
are the smallest and most numerous follicles in the ovary. They contain an immature ovum, surrounded and nourished by a single layer of cells.

Medulla
is the middle part of the ovary that contains blood vessels.

Corpus albicans
is scarlike tissue that is formed when a corpus luteum degenerates.

Fimbriae
are fingerlike extensions that guide the released ovum into the uterine tube.

Primary follicles
develop from some of the primordial follicles. They have layers of cells around the immature ova.

Secondary follicles
are less numerous than primary follicles. They have a fluid-filled cavity that partially surrounds the developing ovum, while some of the outer cells produce estrogen.

Uterine tube
transports the ovum toward the uterus.

Corpus luteum
is formed from the remains of the ruptured follicle. It produces the hormone progesterone, which prepares the uterus to receive a fertilized ovum. It shrivels and degenerates if the ovum is not fertilized.

Ovum (egg)
travels down the uterine tube, waiting to be fertilized.

Mature (Graafian) follicles
are the largest type of follicle, and usually only one forms every month. They contain a maturing ovum within the central fluid-filled cavity.

Ruptured follicle
releases its ovum at ovulation.

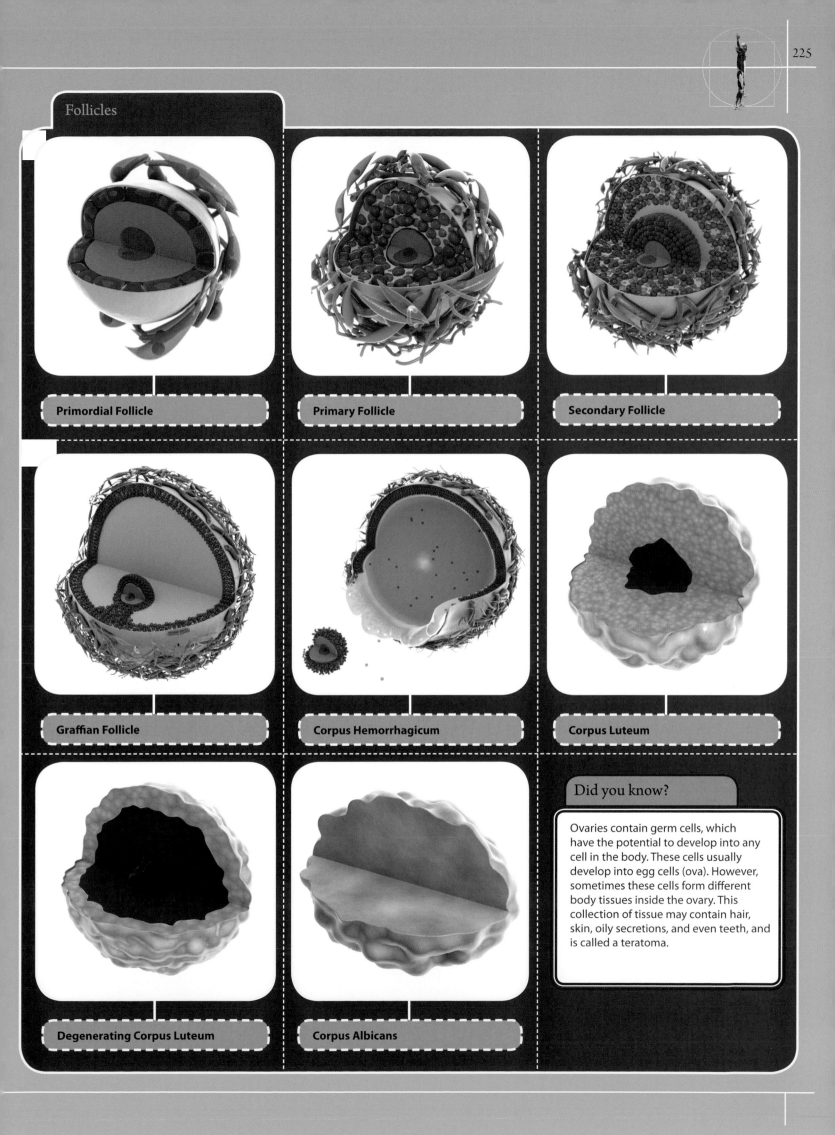

Follicles

Primordial Follicle

Primary Follicle

Secondary Follicle

Graffian Follicle

Corpus Hemorrhagicum

Corpus Luteum

Degenerating Corpus Luteum

Corpus Albicans

Did you know?

Ovaries contain germ cells, which have the potential to develop into any cell in the body. These cells usually develop into egg cells (ova). However, sometimes these cells form different body tissues inside the ovary. This collection of tissue may contain hair, skin, oily secretions, and even teeth, and is called a teratoma.

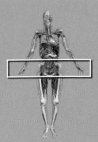

The organs of the female reproductive tract are located in the pelvis. They include the ovaries, uterine (Fallopian) tubes, uterus (womb), and vagina. They produce, store, release, and transport female gametes (ova), providing a place for fertilization to take place, and offering a suitable environment for a fertilized ovum (zygote) to develop and grow.

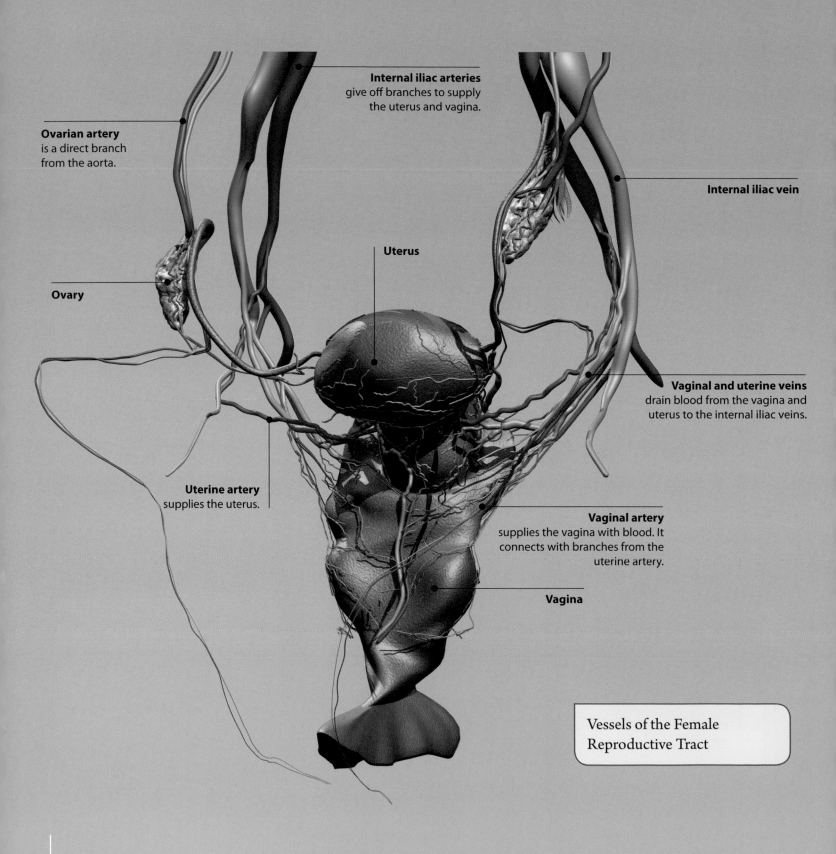

Internal iliac arteries
give off branches to supply the uterus and vagina.

Ovarian artery
is a direct branch from the aorta.

Internal iliac vein

Uterus

Ovary

Vaginal and uterine veins
drain blood from the vagina and uterus to the internal iliac veins.

Uterine artery
supplies the uterus.

Vaginal artery
supplies the vagina with blood. It connects with branches from the uterine artery.

Vagina

Vessels of the Female Reproductive Tract

Female Reproductive Tract: Anterior Aspect

Uterine (Fallopian) tubes transport ova from the ovaries to the uterus. If sperm are present, this is where fertilization takes place.

Ovary are paired organs that produce and store ova (eggs), releasing one each month.

Ligament of the ovary attaches the ovary to the uterus.

Uterus (womb) is a thick-walled muscular chamber where babies can grow and develop. Its lining cells undergo regular changes in response to the hormones estrogen and progesterone.

Vagina is a muscular tube that connects the uterus to the outside, and forms part of the birth canal.

Labia minora are fleshy pieces of tissue that surround the openings of the urethra and vagina.

Clitoris is important in the sexual response.

Infundibulum is the broad end of the uterine tube next to the ovary. Fingerlike projections, called fimbriae, guide the released ova into the uterine tube.

Uterine (Fallopian) tubes

Ovary

Uterus

Cervix is the lower part of the uterus. It forms part of the birth canal.

Vagina

Female Reproductive Tract: Posterior Aspect

Labia minora

The uterus is located in the pelvis, positioned between the urinary bladder and the rectum. The uterus is held in place by a number of ligaments, which attach it to the pelvic bones and surrounding tissues.

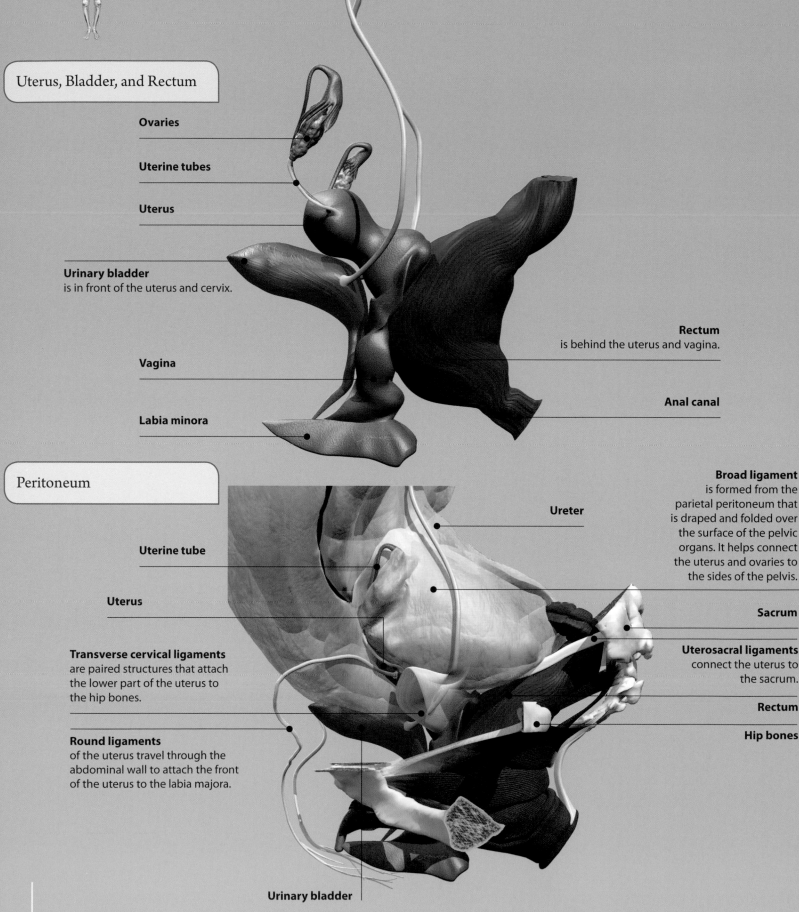

Uterus, Bladder, and Rectum

Ovaries

Uterine tubes

Uterus

Urinary bladder
is in front of the uterus and cervix.

Rectum
is behind the uterus and vagina.

Vagina

Anal canal

Labia minora

Peritoneum

Uterine tube

Uterus

Transverse cervical ligaments
are paired structures that attach the lower part of the uterus to the hip bones.

Round ligaments
of the uterus travel through the abdominal wall to attach the front of the uterus to the labia majora.

Ureter

Broad ligament
is formed from the parietal peritoneum that is draped and folded over the surface of the pelvic organs. It helps connect the uterus and ovaries to the sides of the pelvis.

Sacrum

Uterosacral ligaments
connect the uterus to the sacrum.

Rectum

Hip bones

Urinary bladder

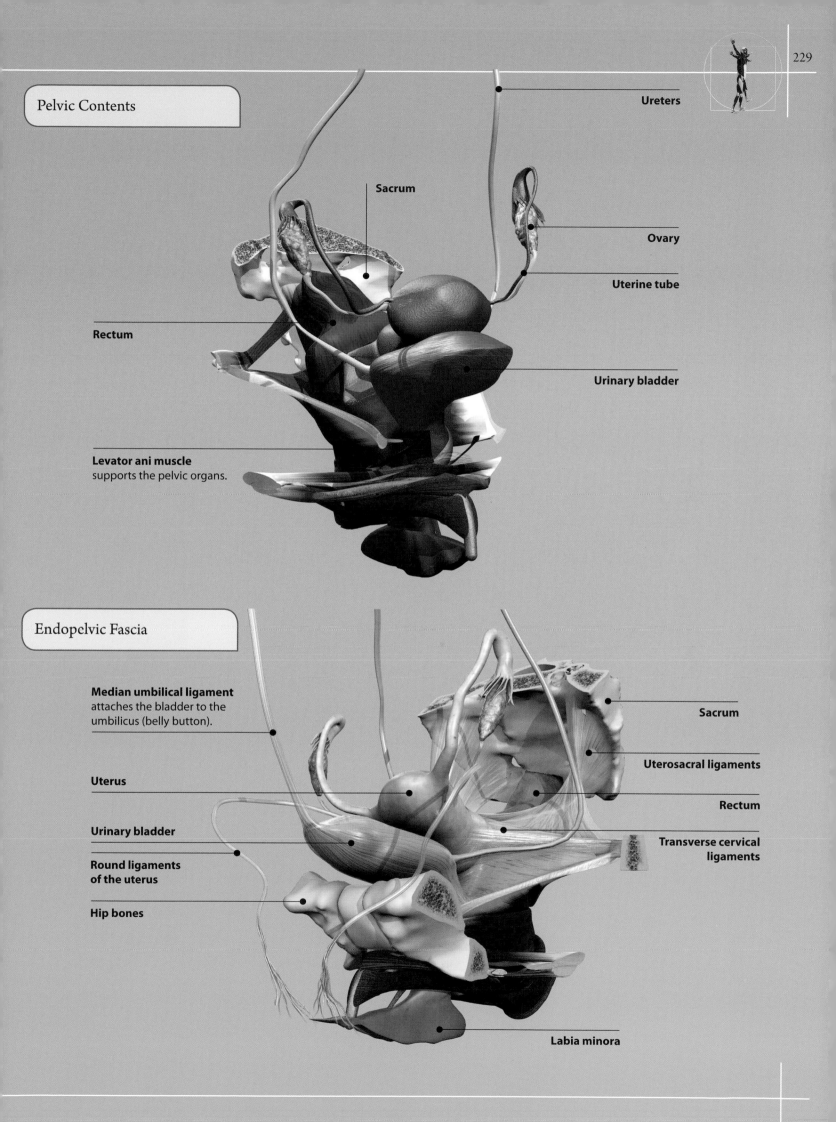

Pelvic Contents

Ureters

Sacrum

Ovary

Uterine tube

Rectum

Urinary bladder

Levator ani muscle
supports the pelvic organs.

Endopelvic Fascia

Median umbilical ligament
attaches the bladder to the
umbilicus (belly button).

Sacrum

Uterosacral ligaments

Uterus

Rectum

Urinary bladder

Transverse cervical ligaments

Round ligaments of the uterus

Hip bones

Labia minora

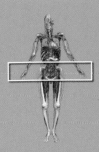

The urinary bladder is a muscular, expandable chamber found in the pelvis. It stores the urine formed by the kidneys until an appropriate place is found to expel it from the body.

The urethra transports urine from the bladder to the outside, in a process called urination.

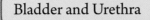

Bladder and Urethra

Urinary bladder
has a thick wall that contains both elastic and muscular fibers. The elastic fibers allow it to stretch and fill with urine, while the muscular fibers contract and squeeze urine down the urethra during urination.

Ureters
are muscular tubes that deliver urine from the kidneys to the bladder.

Right ureteric orifice
is where the right ureter opens into the bladder. It forms a one-way valve to stop urine going back up the ureter. The left ureter enters at the left ureteric orifice.

Trigone
is a smooth triangular area on the inside of the bladder. The triangle is formed by the two ureteric orifices and the internal urethral orifice.

Internal urethral orifice
is the opening where urine leaves the bladder to enter the urethra.

Sphincter urethrae
is a muscle that surrounds the upper urethra. When it is contracted it closes the urethra; when it is relaxed, urine can enter the urethra. This muscle allows us to control the process of urination.

Urethra
is a muscular tube that expels urine out of the body. It is approximately one and a half inches long in females.

External urethral orifice
is the opening of the urethra to the outside.

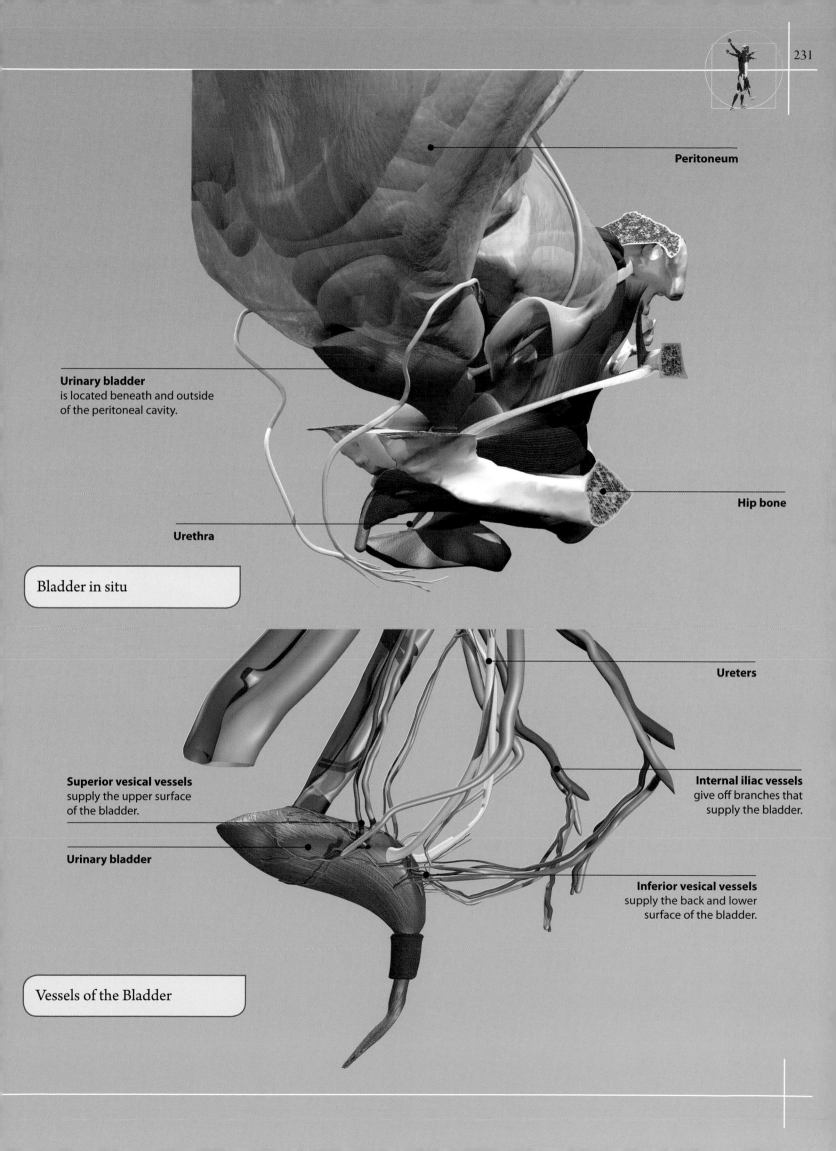

Peritoneum

Urinary bladder
is located beneath and outside
of the peritoneal cavity.

Hip bone

Urethra

Bladder in situ

Ureters

Superior vesical vessels
supply the upper surface
of the bladder.

Internal iliac vessels
give off branches that
supply the bladder.

Urinary bladder

Inferior vesical vessels
supply the back and lower
surface of the bladder.

Vessels of the Bladder

The male reproductive system includes the penis and testes, along with the ducts and glands that connect them. It produces, stores, matures, and delivers male sex cells called sperm. Sperm are suspended in fluids produced by the seminal vesicles and prostate to form semen.

In an intimate act called sexual intercourse, the penis is inserted into the female vagina delivering semen into the female reproductive tract. If sperm meet a female sex cell (ovum) then fertilization takes place, and a new life starts to develop.

The male reproductive system produces hormones such as testosterone, which help with development all over the body.

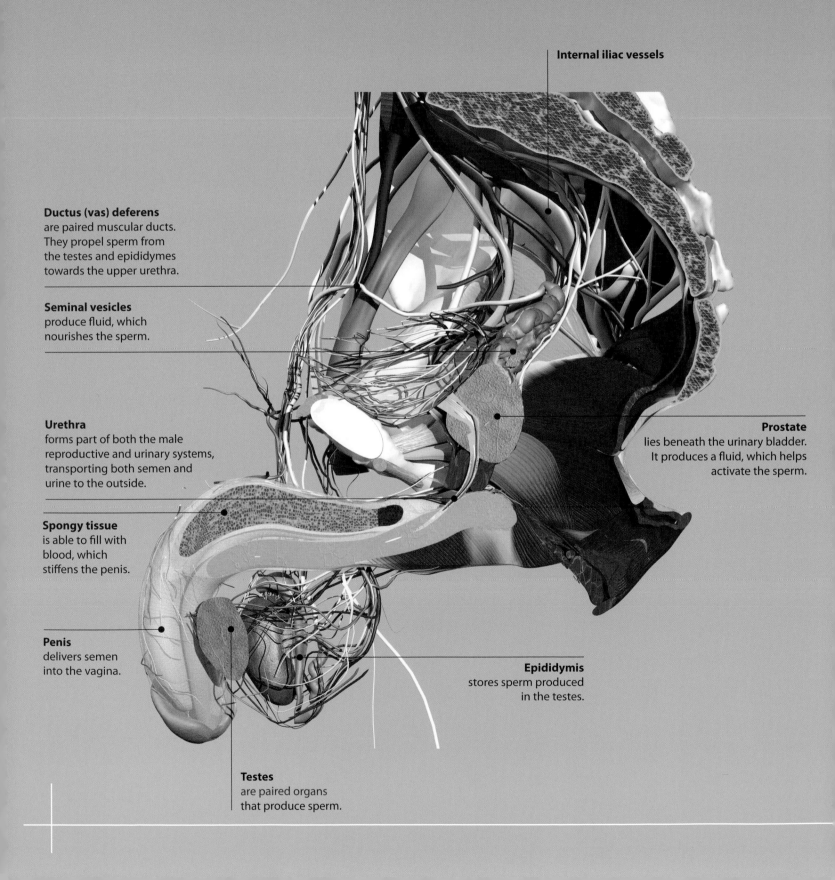

Internal iliac vessels

Ductus (vas) deferens
are paired muscular ducts. They propel sperm from the testes and epididymes towards the upper urethra.

Seminal vesicles
produce fluid, which nourishes the sperm.

Urethra
forms part of both the male reproductive and urinary systems, transporting both semen and urine to the outside.

Spongy tissue
is able to fill with blood, which stiffens the penis.

Penis
delivers semen into the vagina.

Prostate
lies beneath the urinary bladder. It produces a fluid, which helps activate the sperm.

Epididymis
stores sperm produced in the testes.

Testes
are paired organs that produce sperm.

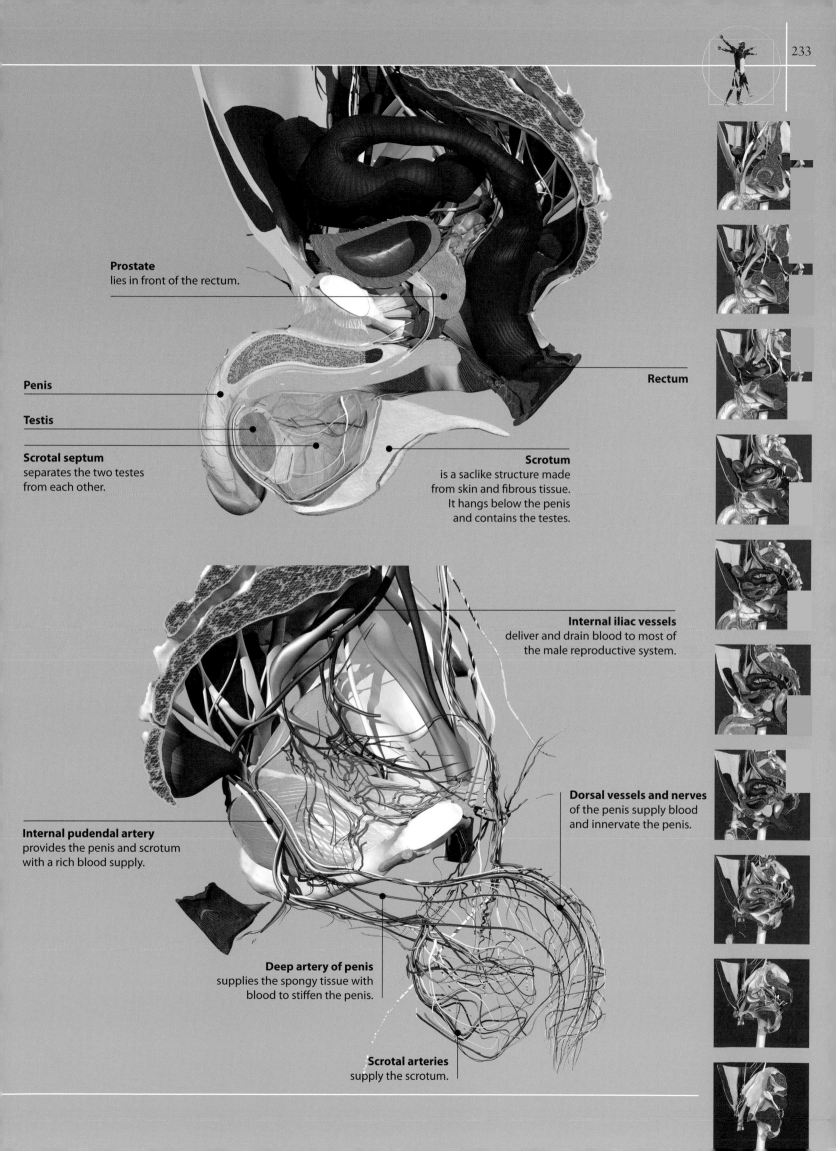

Prostate
lies in front of the rectum.

Penis

Testis

Scrotal septum
separates the two testes
from each other.

Rectum

Scrotum
is a saclike structure made
from skin and fibrous tissue.
It hangs below the penis
and contains the testes.

Internal iliac vessels
deliver and drain blood to most of
the male reproductive system.

Dorsal vessels and nerves
of the penis supply blood
and innervate the penis.

Internal pudendal artery
provides the penis and scrotum
with a rich blood supply.

Deep artery of penis
supplies the spongy tissue with
blood to stiffen the penis.

Scrotal arteries
supply the scrotum.

The pelvic floor is a muscular sheet that forms the base of the pelvic cavity. It is made up of numerous muscles, suspended from the pelvic bones. These form a hammocklike support for the pelvic organs. In the male it has openings for the urethra and anus.

Male Pelvic Floor: Superior Aspect

Sacrum & coccyx
form the back wall of the pelvis.

Coccygeus
is a muscle that forms the back of the pelvic floor, running between the coccyx and the hip bone.

Hip bone
provides attachment for many of the pelvic floor muscles.

Levator ani
is a large muscle, which forms the majority of the pelvic floor.

Internal anal sphincter muscles
surround and compress the upper part of the anal canal.

Sphincter urethrae muscle
surrounds and compresses the urethra, allowing us to control when to urinate.

Opening for anus

Male Pelvic Floor: Inferior Aspect

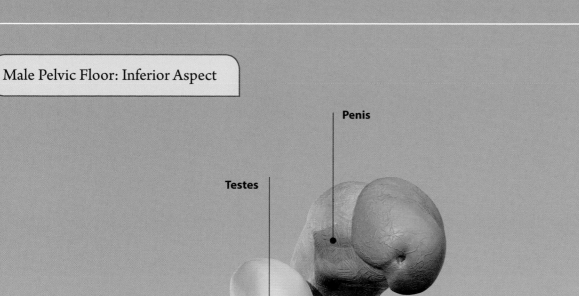

Penis

Testes

Hip bone

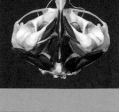

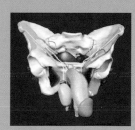

Superficial transverse perineal muscle
attaches the perineal body to the hip bones.

Ischial tuberosity
is the part of the hip
bone that we sit on.

Levator ani

Bulbospongiosus muscle
covers the bulb of the penis.
During sexual intercourse, its
rhythmic contractions squeezes
semen along the urethra.

Ischiocavernosus
attaches to the ischium, and
covers the crura of the penis
on either side.

Anal canal

Coccygeus

External anal sphincter muscle
surrounds and compresses the anal
canal, giving us voluntary control
over the process of defecation.

Perineal body
is a tough fibrous structure lying
in front of the anus. It provides a
central point of attachment for
the pelvic floor muscles.

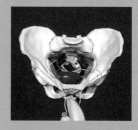

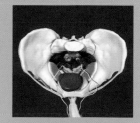

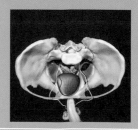

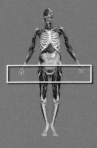

The testes are the paired male reproductive glands. They are located within the scrotum, where they produce male sex cells, called sperm. Mature sperm are stored in the epididymis, until they are released in a process called ejaculation.

The testes originally develop near the kidneys. To reach the scrotum, they descend and pass through the abdominal wall. As they do so, they take with them a sac of peritoneum. This forms a covering around the testes known as the tunica vaginalis.

Efferent ductules
allow sperm to get from the rete testis to the epididymis.

Rete testis
is a branching network of tubes formed from the seminiferous tubules.

Seminiferous tubules
are tightly coiled tubes. They contain germ cells which produce sperm, along with other cells which produce hormones and nourish the developing sperm.

Tunica albuginea
is the tough, white, fibrous tissue capsule around the testes. Bands of tissue penetrate into the testes, dividing it up into lobules.

Lobules
are divisions of the testes that contain the seminiferous tubules. There are about 200-300 lobules in each testis.

Lobules

Epididymis
is found at the back of the testis. It stores the sperm prior to ejaculation.

Ductus deferens
is a muscular tube that transports sperm from the epididymis towards the urethra.

Tunica albuginea

Tunica vaginalis
is formed from the peritoneum during the development and descent of the testes. It forms a double–layered covering around each testis.

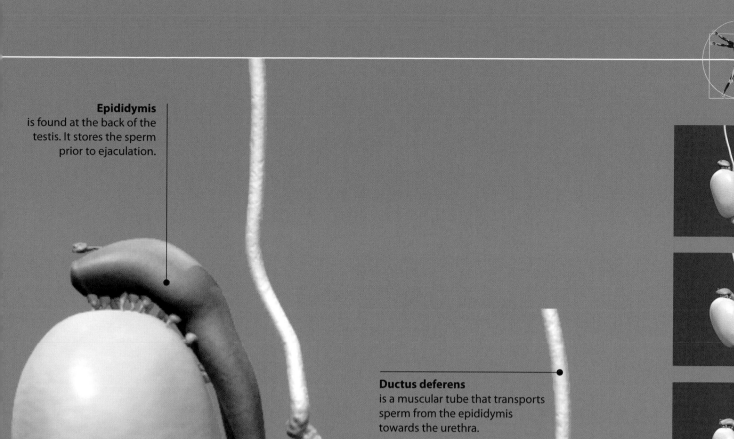

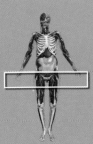

Spermatogenesis is the process by which the male sex cells, called sperm, are formed. It occurs in the lining cells of the tightly coiled seminiferous tubules, located within the testes. After puberty, approximately 300 million sperm are produced in the testes each day.

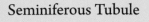

Seminiferous Tubule

Seminiferous tubules
contain germ cells which mature to form sperm, along with their supporting Sertoli cells.

Basement membrane
supports the seminiferous epithelium.

Lumen
is the fluid-filled, hollow, inner part of the seminiferous tubules. It contains fluid and transports sperm to the rete testis.

Seminiferous epithelium
lines the inside of the seminiferous tubules. It is made up of Sertoli cells which surround the developing sperm.

Sertoli cells
surround developing sperm, supporting and nourishing them. They also produce hormones, and a protein-rich fluid, which is secreted into the lumen.

Sertoli Cell

Late spermatids
have a more developed acrosome
and tail than early spermatids.

Early spermatids
are more arrow-shaped,
with a tail starting to form.

Spermatogonia
are the germ cells which divide
to provide a constant supply of
developing male sex cells (sperm).
They are found on top of the
basement membrane.

Sperm
are released into the
seminiferous tubule
lumen in a process
called spermiation.

Secondary spermatocyte
are formed from primary
spermatocytes.

Primary spermatocytes
are formed from
spermatogonia, and are
found closer to the lumen.

Sertoli cell

Basement membrane

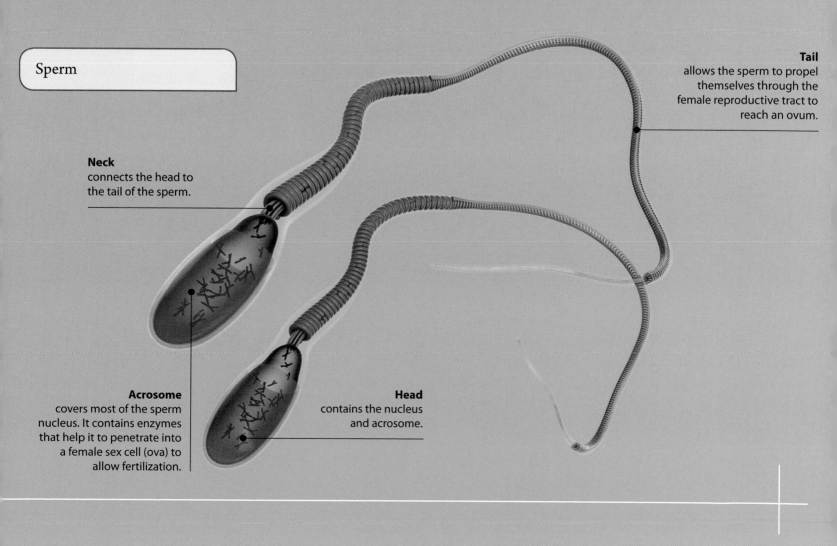

Sperm

Neck
connects the head to
the tail of the sperm.

Tail
allows the sperm to propel
themselves through the
female reproductive tract to
reach an ovum.

Acrosome
covers most of the sperm
nucleus. It contains enzymes
that help it to penetrate into
a female sex cell (ova) to
allow fertilization.

Head
contains the nucleus
and acrosome.

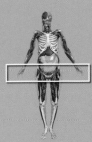

Spermatogenesis occurs best at temperatures slightly lower than body temperature. Due to this, the testes are suspended outside the body within the thin-skinned, saclike scrotum.

During development, as the testes descend from the abdomen into the scrotum, they pass through the abdominal wall. As they do so, they are covered by three layers of tissue called fascia. The testicular blood vessels, and ductus deferens are also lined by these fascia layers to form the spermatic cords.

The short angled passageway through the abdominal wall is called the inguinal canal.

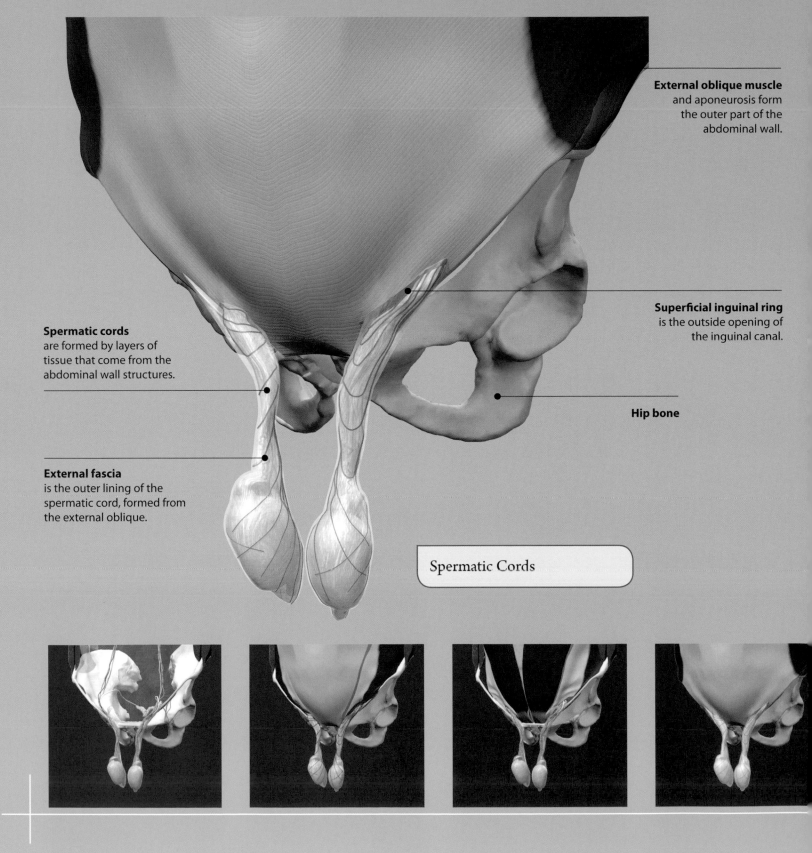

External oblique muscle and aponeurosis form the outer part of the abdominal wall.

Superficial inguinal ring is the outside opening of the inguinal canal.

Spermatic cords are formed by layers of tissue that come from the abdominal wall structures.

Hip bone

External fascia is the outer lining of the spermatic cord, formed from the external oblique.

Spermatic Cords

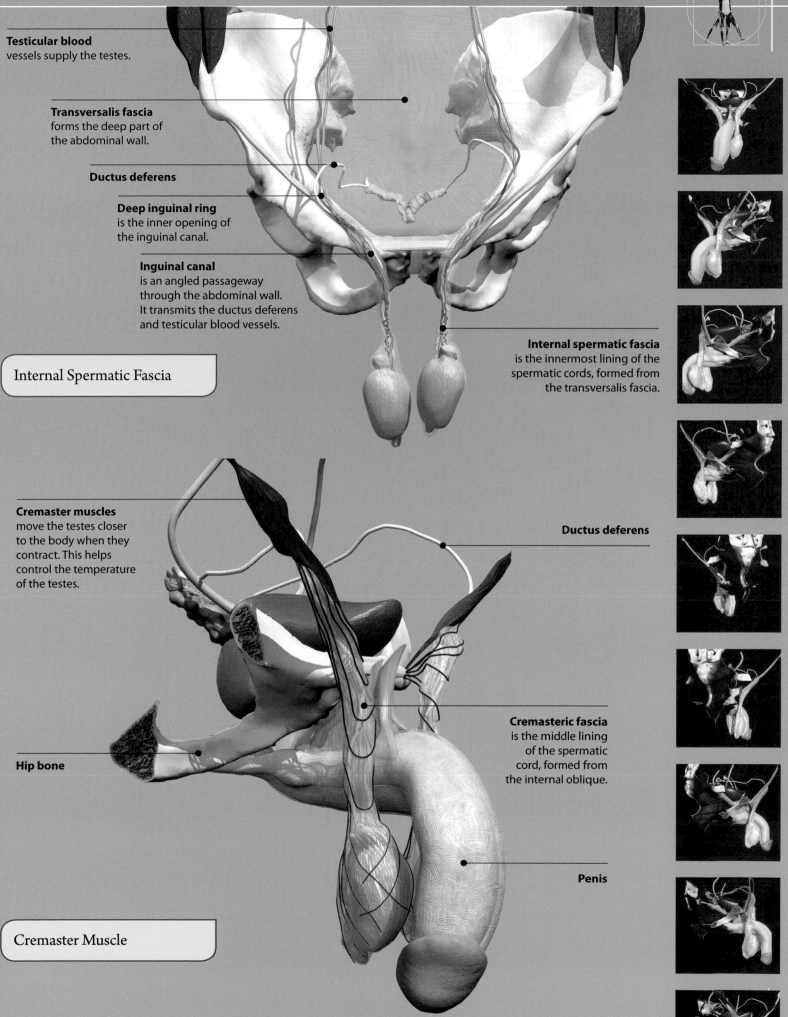

Testicular blood vessels supply the testes.

Transversalis fascia forms the deep part of the abdominal wall.

Ductus deferens

Deep inguinal ring is the inner opening of the inguinal canal.

Inguinal canal is an angled passageway through the abdominal wall. It transmits the ductus deferens and testicular blood vessels.

Internal spermatic fascia is the innermost lining of the spermatic cords, formed from the transversalis fascia.

Internal Spermatic Fascia

Cremaster muscles move the testes closer to the body when they contract. This helps control the temperature of the testes.

Ductus deferens

Hip bone

Cremasteric fascia is the middle lining of the spermatic cord, formed from the internal oblique.

Penis

Cremaster Muscle

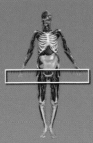

The epididymis, ductus deferens, and ejaculatory ducts are responsible for the storage and transport of sperm from the testes to the urethra.

The accessory glands consist of the seminal vesicles, prostate gland, and bulbourethral glands. Together, they produce seminal fluid, which nourishes the sperm. The mixture of sperm and seminal fluid is known as semen.

Seminal Vesicles

Ampulla
is the widened, twisted end of the ductus deferens, just before it meets the seminal vesicles.

Ureters

Seminal vesicles
are irregularly shaped, blind-ending sacs found behind the urinary bladder. They produce the majority of seminal fluid, which helps nourish and support the sperm. The seminal vesicle ducts join with the ductus deferens to form the ejaculatory ducts.

Ductus deferens

Urinary bladder

Ductus Deferens

Ductus deferens
is a long, thin, muscular tube that transports sperm from the epididymis to the ejaculatory ducts. It travels in the spermatic cord and through the inguinal canal, before joining with ducts draining the seminal vesicles to form the ejaculatory ducts.

Prostate gland
is found just beneath the male urinary bladder. It produces prostatic fluid, which helps improve the mobility and survival of sperm. The ejaculatory ducts and part of the urethra run through the prostate gland.

Ureters

Penis

Bulbourethral glands
are found at the base of the penis. They produce a natural lubricating fluid.

Seminal vesicles

Epididymes

Urinary bladder

Prostate gland

Testis

Epididymis

Testis

Penis

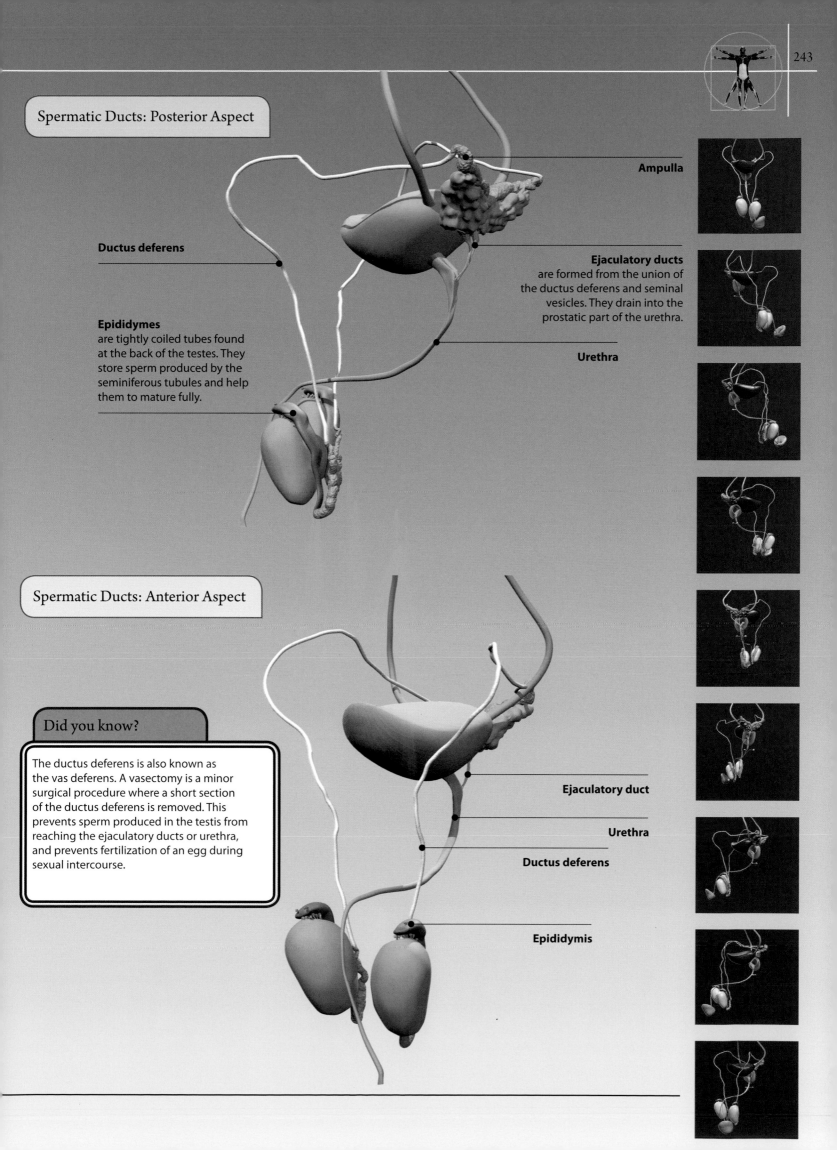

Spermatic Ducts: Posterior Aspect

Ampulla

Ductus deferens

Ejaculatory ducts
are formed from the union of
the ductus deferens and seminal
vesicles. They drain into the
prostatic part of the urethra.

Epididymes
are tightly coiled tubes found
at the back of the testes. They
store sperm produced by the
seminiferous tubules and help
them to mature fully.

Urethra

Spermatic Ducts: Anterior Aspect

Did you know?

The ductus deferens is also known as
the vas deferens. A vasectomy is a minor
surgical procedure where a short section
of the ductus deferens is removed. This
prevents sperm produced in the testis from
reaching the ejaculatory ducts or urethra,
and prevents fertilization of an egg during
sexual intercourse.

Ejaculatory duct

Urethra

Ductus deferens

Epididymis

The penis is a cylindrical organ that allows semen to be delivered into the vagina during sexual intercourse, and also provides an outlet for urine. It is made up of three tubular structures (corpora), with the urethra running along its length. The penis can be made stiff by filling the corpora with blood, leading to an erection.

Hip bone
provides attachment for the crura of the penis.

Fundiform and suspensory ligaments
of the penis help support the weight of the penis.

Body
is the shaft of the penis containing the three corpora.

Ductus deferens

Epididymis

Ischiocavernosus muscles
cover the crura of the penis. They are important in helping to keep the penis stiff during an erection.

Superficial penile fascia
covers the three corpora of the penis, keeping them together.

Bulbospongiosus muscle
covers the bulb of the penis. Rhythmic contractions of this muscle propel semen along the urethra during ejaculation.

Testes

Root
of the penis attaches it to the body.

Glans
is the widened end of the penis where the urethra opens.

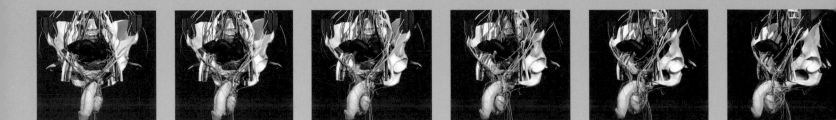

Urethra
runs through the
corpus spongiosum.

Crus
attach the corpus cavernosum
on each side to the hip bone.

Bulb
is the expanded root of
the corpus spongiosum.

Corpora cavernosa
are paired tubular
structures that are filled
with spongy erectile tissue.

Corpus spongiosum
contains the urethra. It
expands at the end of the
penis to form the glans.

Glans penis
is the expanded end of
the corpus spongiosum. In
uncircumcised males, it is
covered by the foreskin.

Internal pudendal blood vessels
supply and drain the arteries and
veins of the penis.

**Arteries and veins
of the penis**

Urethra

Corpus spongiosum

Tunica albuginea
is a white fibroelastic
layer of tissue, covering
and connecting the
corpora cavernosa.

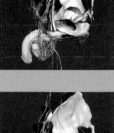
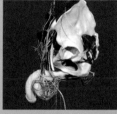
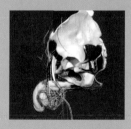
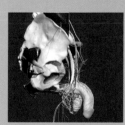
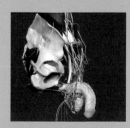
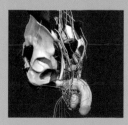
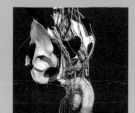

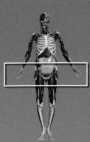

The urinary bladder is a muscular, expandable chamber found in the pelvis. It stores the urine formed by the kidneys until an appropriate place is found to expel it from the body.

The urethra is a muscular tube that connects the urinary bladder to the outside. In the male it is an outlet for both urine and semen, and is much longer than the female urethra (approximately eight inches long). It is divided into three sections: prostatic, membranous, and spongy (penile).

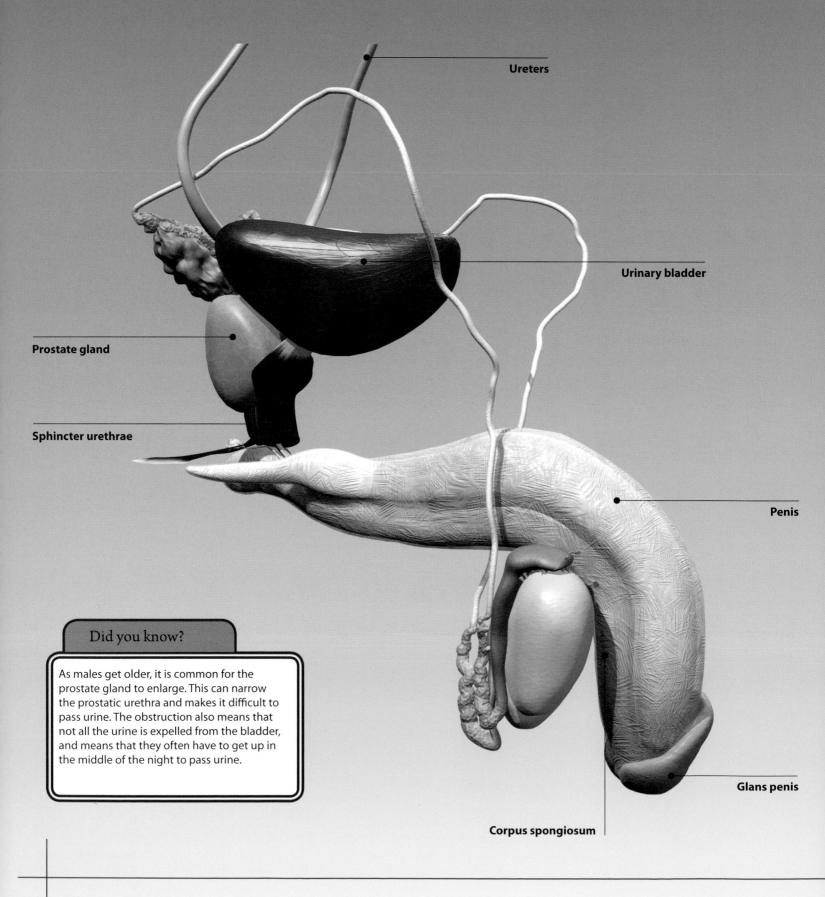

Ureters

Urinary bladder

Prostate gland

Sphincter urethrae

Penis

Did you know?

As males get older, it is common for the prostate gland to enlarge. This can narrow the prostatic urethra and makes it difficult to pass urine. The obstruction also means that not all the urine is expelled from the bladder, and means that they often have to get up in the middle of the night to pass urine.

Glans penis

Corpus spongiosum

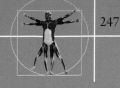

Ureters
are muscular tubes that deliver urine from the kidneys to the bladder.

Urinary bladder
stores urine formed by the kidneys.

Internal urethral orifice
is the opening where urine leaves the bladder to enter the urethra.

Prostatic urethra
is the portion of the urethra that runs through the prostate gland, and is approximately one inch long. The ejaculatory ducts open into it.

Prostate gland
lies beneath the urinary bladder.

Membranous urethra
is the short section of urethra approximatley half an inch long that runs through the pelvic floor muscles.

Sphincter urethrae
is a muscle that surrounds the membranous urethra and gives us control over the process of urination. When it is contracted it closes the urethra; when it is relaxed, urine can enter the urethra.

Spongy (penile) urethra
is the longest section of the male urethra approximatley six inches long and runs within the corpus spongiosum until it reaches the external urethral orifice.

Corpus spongiosum
contains the spongy (penile) urethra.

External urethral orifice
is the opening of the urethra to the outside at the glans penis.

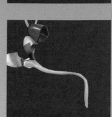

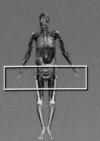

The rectum and anal canal are the final sections of the digestive system. The rectum is located in the pelvis in front of the sacrum, and is a continuation of the sigmoid colon. It is about five inches long, and passes through the pelvic floor muscles to become the anal canal. The rectum and anal canal are involved in the process of defecation, where feces are expelled from the body.

Coronal Section Through the Rectum

Inferior mesenteric artery
supplies the sigmoid colon and part of the rectum.

Sigmoid colon
becomes the rectum in front of the sacrum.

Hip bone

Sacrum

Ischiorectal fossae
are fat-filled spaces found either side of the anal canal beneath the pelvic floor.

Rectum
is found in front of the sacrum. A buildup of feces inside the rectum triggers the urge to defecate. External anal sphincter surrounds the anal canal. When it contracts it closes the anal canal; when it relaxes the anal canal is opened and feces can be expelled.

Pelvic floor
is a muscular hammock-shaped structure that supports the pelvic organs. The rectum passes through an opening in the pelvic floor to become the anal canal.

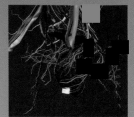
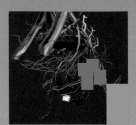
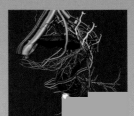

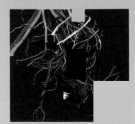

Blood Supply to the Rectum

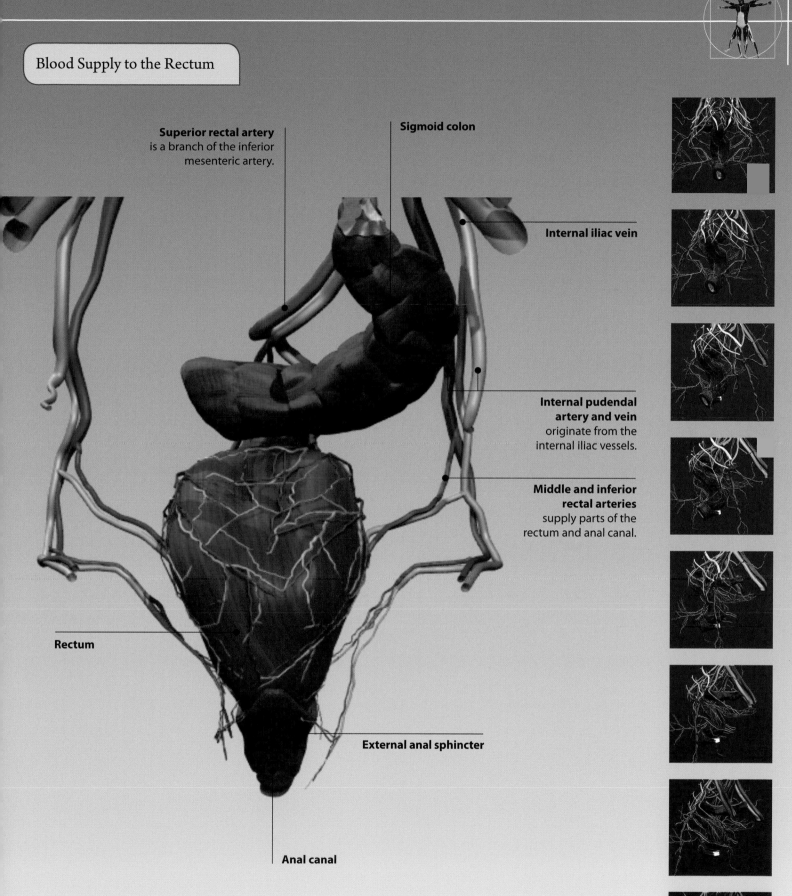

Superior rectal artery is a branch of the inferior mesenteric artery.

Sigmoid colon

Internal iliac vein

Internal pudendal artery and vein originate from the internal iliac vessels.

Middle and inferior rectal arteries supply parts of the rectum and anal canal.

Rectum

External anal sphincter

Anal canal

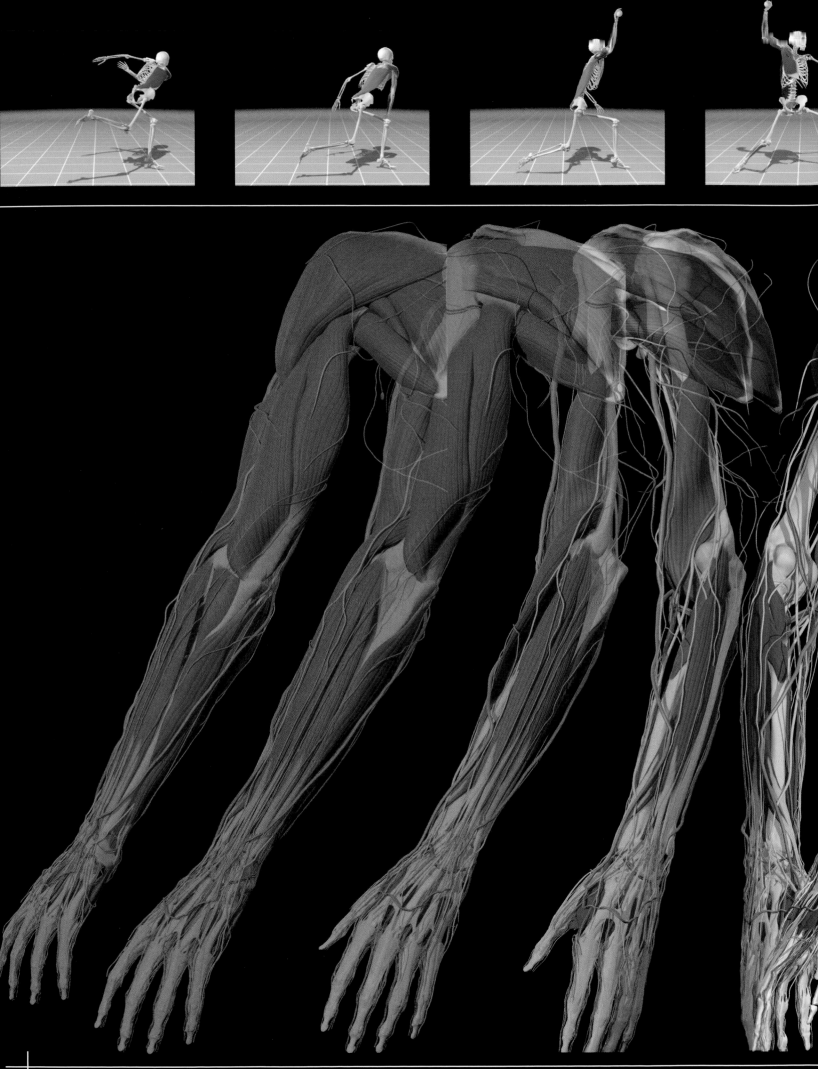

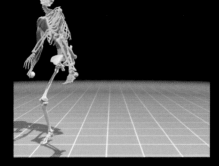

The two upper limbs are attached to the trunk by the shoulder girdle. This is formed by the collar bones (clavicle) and shoulder blades (scapula). Each upper limb can be divided into four main regions: the shoulder, arm, forearm, and hand.

Movements of the shoulder and elbow joints position the hand precisely in space, so that it can carry out a wide range of functions. These vary from the delicate, coordinated actions required to tie shoe laces, through to gripping, lifting, moving, and throwing objects. The hand also has numerous sensory receptors, which allow us to tell the difference between objects just on the basis of touch.

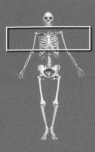

The shoulder is an example of a ball and socket joint. It is formed by the rounded head of the humerus (arm bone) moving within the glenoid cavity of the scapula (shoulder blade). It is the most mobile joint in the body, capable of a wide range of movements. Strong ligaments and muscles help to make the joint stable.

Bones of the Shoulder Joint

Acromion
is a bony projection from the scapula that forms the "tip" of the shoulder.

Glenoid cavity
is a shallow depression on the side of the shoulder blade (scapula).

Head of humerus
is smooth and rounded. It moves within the shallow glenoid cavity of the scapula (shoulder blade).

Coracoid process
is a bony projection, which provides attachments for various ligaments around the shoulder.

Scapula, or shoulder blade,
forms part of the shoulder joint, as well as providing attachments for the surrounding ligaments and muscles.

Humerus, or arm bone.
Its upper end contributes to the shoulder joint.

Ligaments of the Shoulder Joint

Coracoclavicular ligament

Acromioclavicular ligament

Coracoacromial ligament

Transverse humeral ligament

Biceps brachii tendon

First rib

Glenohumeral ligaments reinforce the joint capsule.

Clavicle (or collar bone) forms part of the shoulder girdle, attaching the upper limb to the trunk.

Coronal Section Through the Shoulder Joint

Ligaments help protect and stabilize the shoulder joint.

Clavicle

Scapula

Acromion

Biceps brachii tendon runs through the shoulder joint to attach to the scapula.

Head of humerus

Glenoid cavity

Glenoid labrum is a ring of tough cartilage that runs around the edge of the glenoid cavity, making it deeper.

Rotator cuff muscles move the shoulder joint, and help stabilize it.

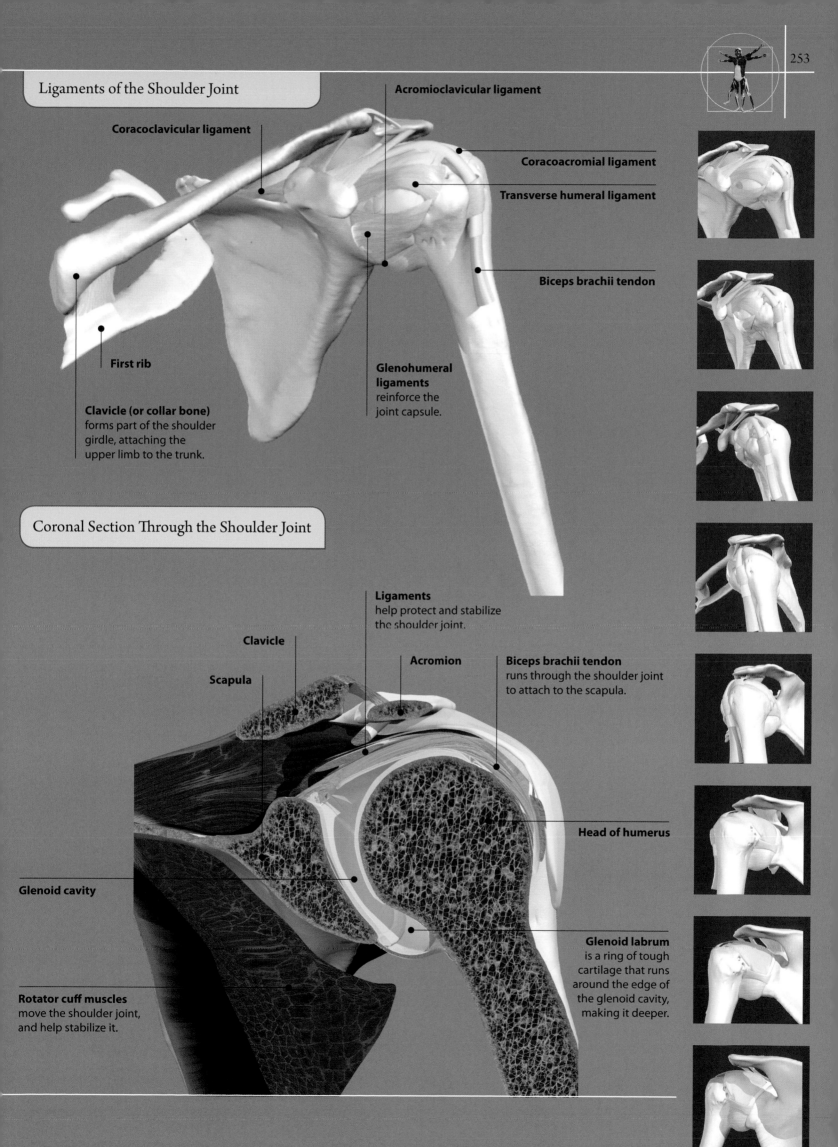

The shoulder is the region of the upper limb that attaches it to the trunk. It contains the highly mobile shoulder joint. Muscles acting at this joint move the upper limb forward and backward, out to the side, and back across the body, as well as turning it inwards and outwards.

The rotator cuff is formed by four muscles that are attached to the scapula, and humerus. As well as producing movements at the shoulder, they also stabilize the shoulder joint, and prevent it from dislocating (popping out). The rotator cuff muscles are: supraspinatus, infraspinatus, teres minor, and subscapularis.

Tendonitis

Acromion

Inflamed bursa

Supraspinatus

The tendon of the supraspinatus muscle passes beneath the acromion, through the narrow subacromial space. A fluid-lined sac, called the subacromial bursa, helps prevent friction from repetitive movements of the tendon. However, sometimes the bursa and tendon can get irritated and inflamed. This leads to pain when moving the arm out to the side, and is known as subacromial bursitis or supraspinatus tendonitis.

Did you know?

The deltoid muscle is a common site for giving intramuscular injections, due to its large size and easy accessibility.

Deep Muscles: Anterior Aspect

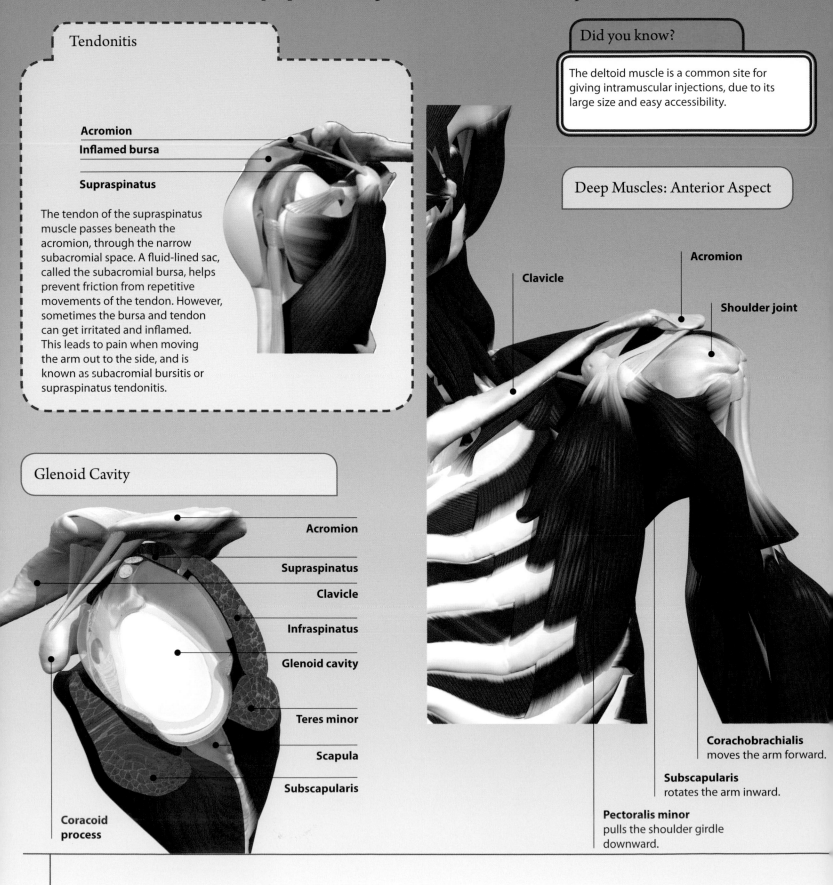

Clavicle

Acromion

Shoulder joint

Glenoid Cavity

Acromion

Supraspinatus

Clavicle

Infraspinatus

Glenoid cavity

Teres minor

Scapula

Subscapularis

Coracoid process

Corachobrachialis moves the arm forward.

Subscapularis rotates the arm inward.

Pectoralis minor pulls the shoulder girdle downward.

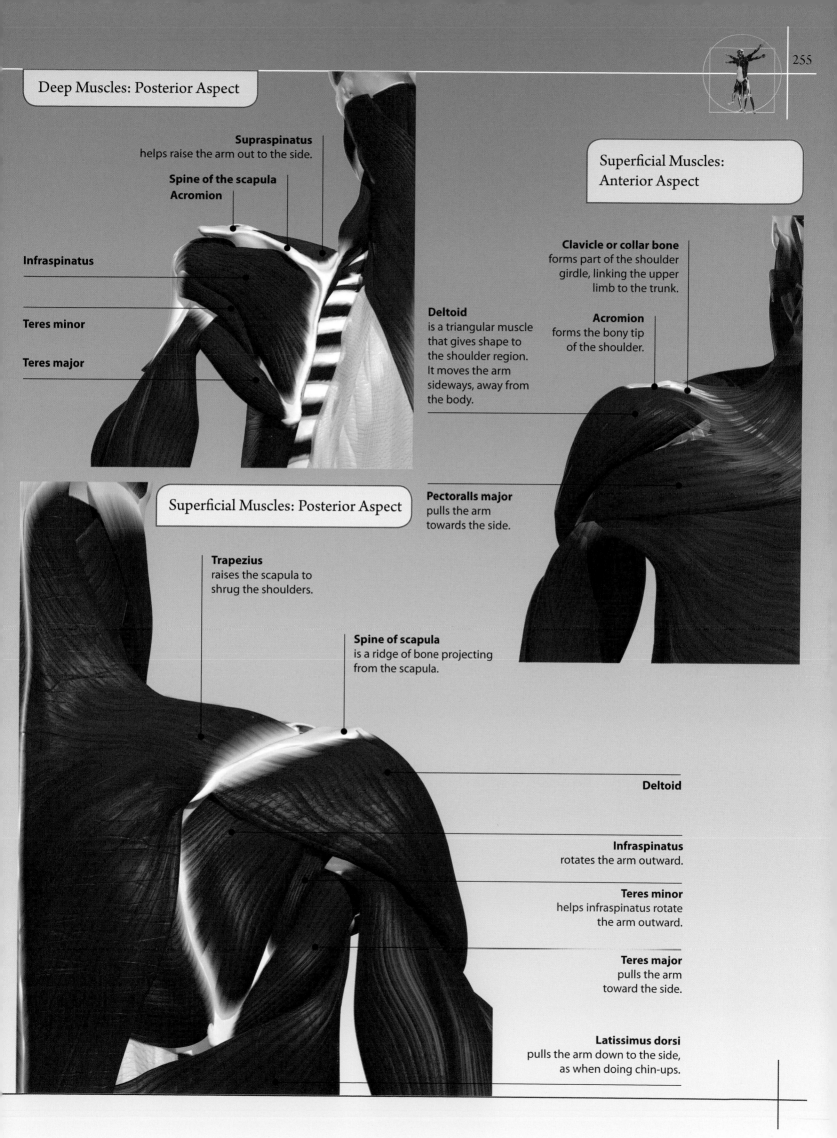

Deep Muscles: Posterior Aspect

Supraspinatus
helps raise the arm out to the side.

Spine of the scapula
Acromion

Infraspinatus

Teres minor

Teres major

Superficial Muscles: Anterior Aspect

Clavicle or collar bone
forms part of the shoulder girdle, linking the upper limb to the trunk.

Deltoid
is a triangular muscle that gives shape to the shoulder region. It moves the arm sideways, away from the body.

Acromion
forms the bony tip of the shoulder.

Superficial Muscles: Posterior Aspect

Pectoralls major
pulls the arm towards the side.

Trapezius
raises the scapula to shrug the shoulders.

Spine of scapula
is a ridge of bone projecting from the scapula.

Deltoid

Infraspinatus
rotates the arm outward.

Teres minor
helps infraspinatus rotate the arm outward.

Teres major
pulls the arm toward the side.

Latissimus dorsi
pulls the arm down to the side, as when doing chin-ups.

The brachial plexus is a network of nerves formed in the lower part of the neck. It is made up of roots, trunks, divisions, cords, and branches. The branches of the brachial plexus supply sensation and motor function to the entire upper limb. The five main branches of the brachial plexus are the axillary, median, musculocutaneous, radial, and ulnar nerves.

Spinal cord
carries nerve fibers to and from the brain.

Scapula

Scalene muscles
run from the cervical vertebrae to the first rib. The roots of the brachial plexus emerge between them.

Omohyoid
runs over the top of the brachial plexus in the lower part of the neck.

Humerus

Corachobrachialis muscle
is supplied and pierced by the musculocutaneous nerve.

Musculocutaneous nerve

Subclavian artery and vein

Trunks
are formed from the roots.

Divisions
are formed as the trunks split into front and back fibers.

Cords
are closely related to the axillary artery.

Roots of the brachial plexus
come from the 5th to 8th cervical nerves, and 1st thoracic nerve.

Axillary artery

Subclavian artery

Branches
supply the upper limb.

Musculocutaneous nerve

Brachial artery

Median nerve

Ulnar nerve

Nerves of the Brachial Plexus

Median nerve
innervates muscles of the forearm that bend the fingers and wrist. It also supplies sensation to the thumb, index, and middle finger.

Musculocutaneous nerve
innervates muscles of the arm that bend the elbow. It also supplies sensation to part of the forearm.

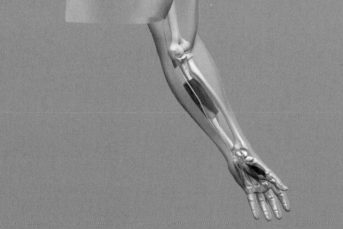

Ulnar nerve
innervates the small muscles of the hand along with a few muscles of the forearm. It supplies sensation to the skin over the ring and little finger.

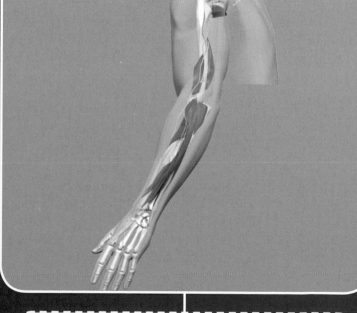

Radial nerve
innervates muscles of the arm and forearm that straighten the fingers, wrist, and elbow. It also supplies sensation to the back of the arm, forearm, and part of the hand.

The shoulder region is packed with blood vessels and nerves, which pass through the axilla (armpit) to reach the rest of the upper limb.

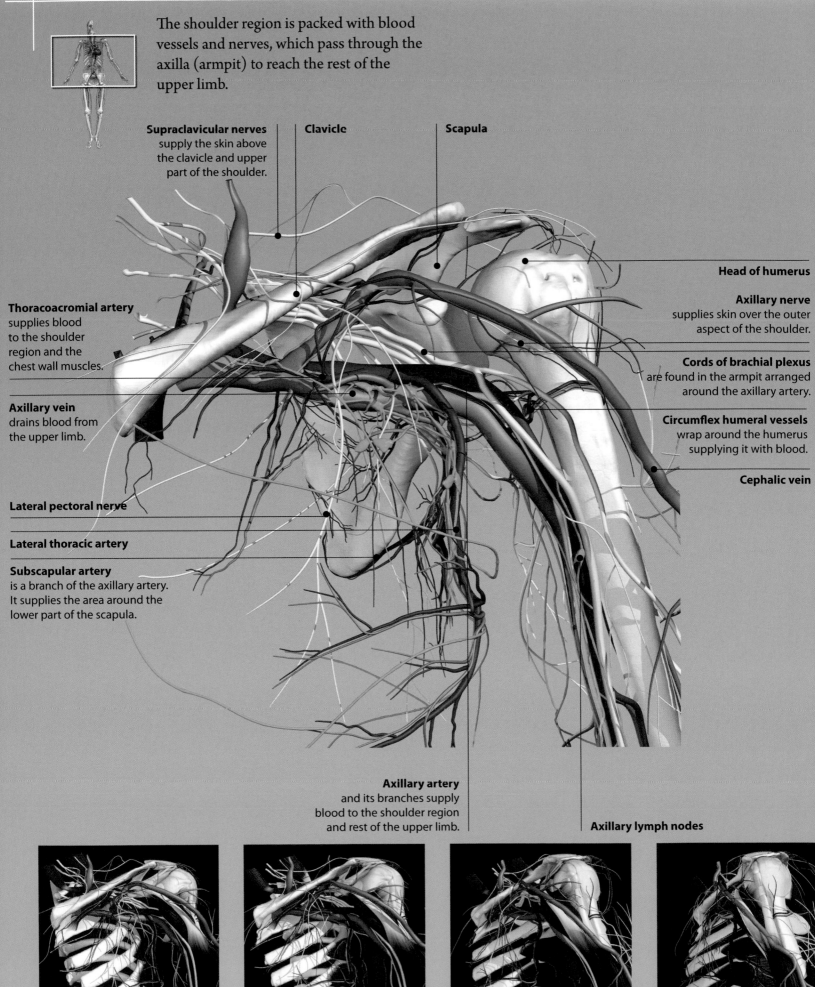

Supraclavicular nerves supply the skin above the clavicle and upper part of the shoulder.

Clavicle

Scapula

Head of humerus

Axillary nerve supplies skin over the outer aspect of the shoulder.

Thoracoacromial artery supplies blood to the shoulder region and the chest wall muscles.

Cords of brachial plexus are found in the armpit arranged around the axillary artery.

Axillary vein drains blood from the upper limb.

Circumflex humeral vessels wrap around the humerus supplying it with blood.

Cephalic vein

Lateral pectoral nerve

Lateral thoracic artery

Subscapular artery is a branch of the axillary artery. It supplies the area around the lower part of the scapula.

Axillary artery and its branches supply blood to the shoulder region and rest of the upper limb.

Axillary lymph nodes

Neurovascular Structures and Deep Muscles

Clavicle

Acromion

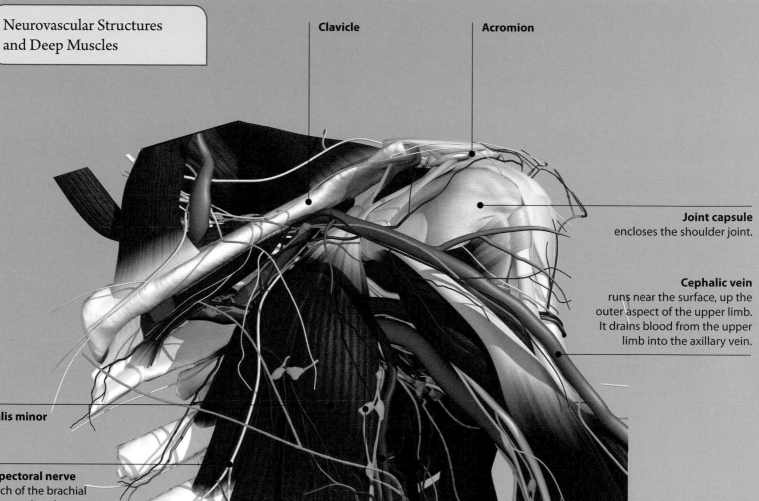

Joint capsule
encloses the shoulder joint.

Cephalic vein
runs near the surface, up the outer aspect of the upper limb. It drains blood from the upper limb into the axillary vein.

Pectoralis minor

Lateral pectoral nerve
is a branch of the brachial plexus. It supplies the pectoralis major muscle.

Axillary lymph nodes
receive tissue fluid (lymph) from the upper limb and part of the chest wall.

Lateral thoracic artery
supplies blood to the outer aspect of the chest wall.

Subscapular artery

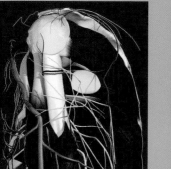
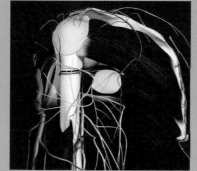
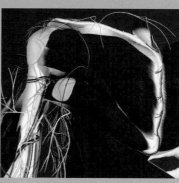
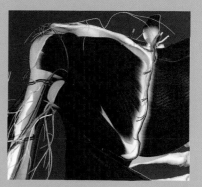

The arm is the part of the upper limb between the shoulder and elbow. The humerus is the arm bone. It provides attachment for various muscles that act at the elbow joint.

As well as flexing (bending) and straightening (extending) the elbow, some muscles spanning the elbow are also able to rotate the forearm. These specialized movements are called pronation (moving the palm of the hand inwards) and supination (moving the palm of the hand outwards).

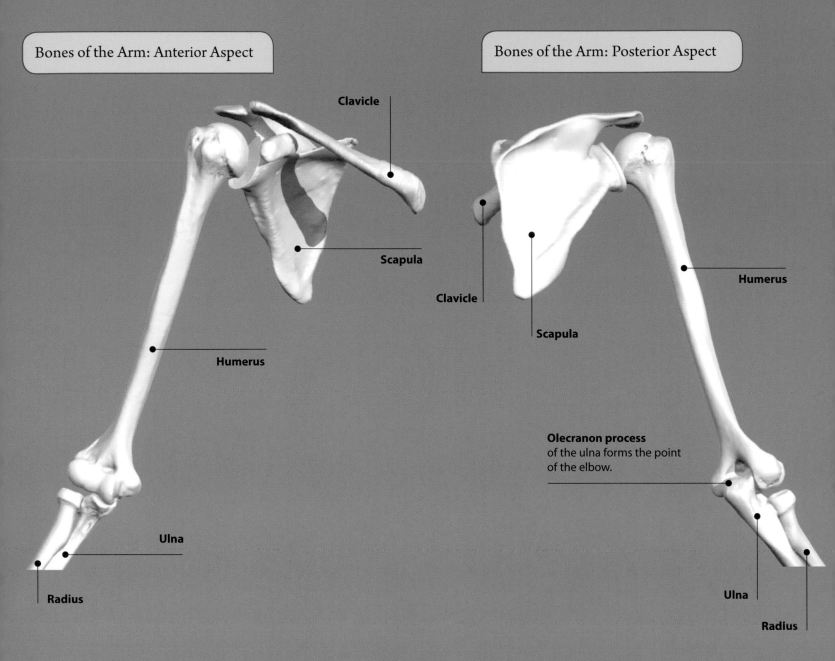

Bones of the Arm: Anterior Aspect

Clavicle

Scapula

Humerus

Ulna

Radius

Bones of the Arm: Posterior Aspect

Clavicle

Scapula

Humerus

Olecranon process
of the ulna forms the point of the elbow.

Ulna

Radius

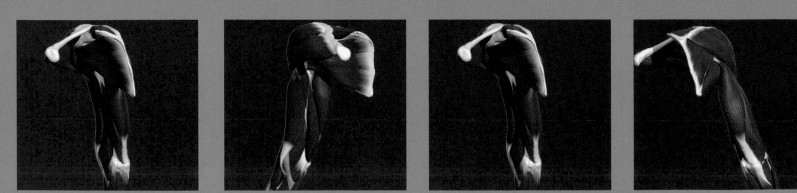

   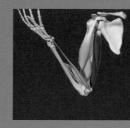 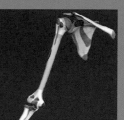 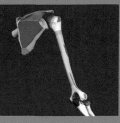

Deep Muscles of the Arm: Anterior Aspect

Deep Muscles of the Arm: Posterior Aspect

Scapula

Clavicle

Tendon of long head of biceps brachii
runs through the shoulder joint capsule, to attach to the scapula at the top of the glenoid cavity.

Tendon of short head of biceps brachii
attaches to the coracoid process of the scapula.

Humerus

Biceps brachii
is a large muscle on the front of the arm. It is attached to the scapula and radius. It is flexes the elbow, as well as supinating the forearm.

Brachialis
muscle lies beneath biceps brachii. It is attached to the humerus and ulna, and is a strong flexor of the elbow.

Pronator teres
is a forearm muscle which pronates the forearm, turning the palm of the hand inward.

Brachioradialis
is a forearm muscle which helps flex the elbow.

Scapula

Humerus

Biceps brachii

Triceps
is a large muscle at the back of the humerus, which extends the elbow. Its three heads are attached to the scapula and humerus, and insert onto the olecranon process.

Brachialis

Anconeus
is a small muscle that helps straighten the elbow joint.

  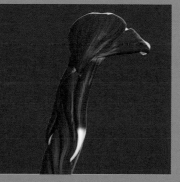 

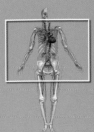

The brachial artery and its branches supply blood to the arm. The cephalic and basilic veins run through the arm, draining blood from the upper limb. Branches from the brachial plexus supply muscles and skin within the upper limb.

Arteries of the Arm: Posterior Aspect

Subclavian artery arises within the thorax. It supplies the upper limb, becoming the axillary artery after it crosses the first rib.

Nerves of the Arm: Posterior Aspect

Axillary nerve

Median nerve

Radial nerve wraps around the humerus.

Musculocutaneous nerve

Axillary artery is a continuation of the subclavian artery. It gives off branches to the shoulder and chest wall.

Profunda brachii artery is a large branch of the brachial artery.

Brachial artery

Radial collateral artery

Ulnar nerve runs behind the bottom of the humerus. This nerve creates the odd sensation when we strike our "funny bone."

Superior ulnar collateral artery

Subclavian vein

Did you know?

The pulsations of the brachial artery can be felt on the inside of the arm by pressing it against the humerus.

Axillary vein

Axillary lymph nodes receive tissue fluid (lymph) from the upper limb.

Cephalic vein

Basilic vein

Veins of theArm: Posterior Aspect

Brachial veins accompany the brachial artery.

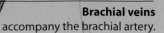

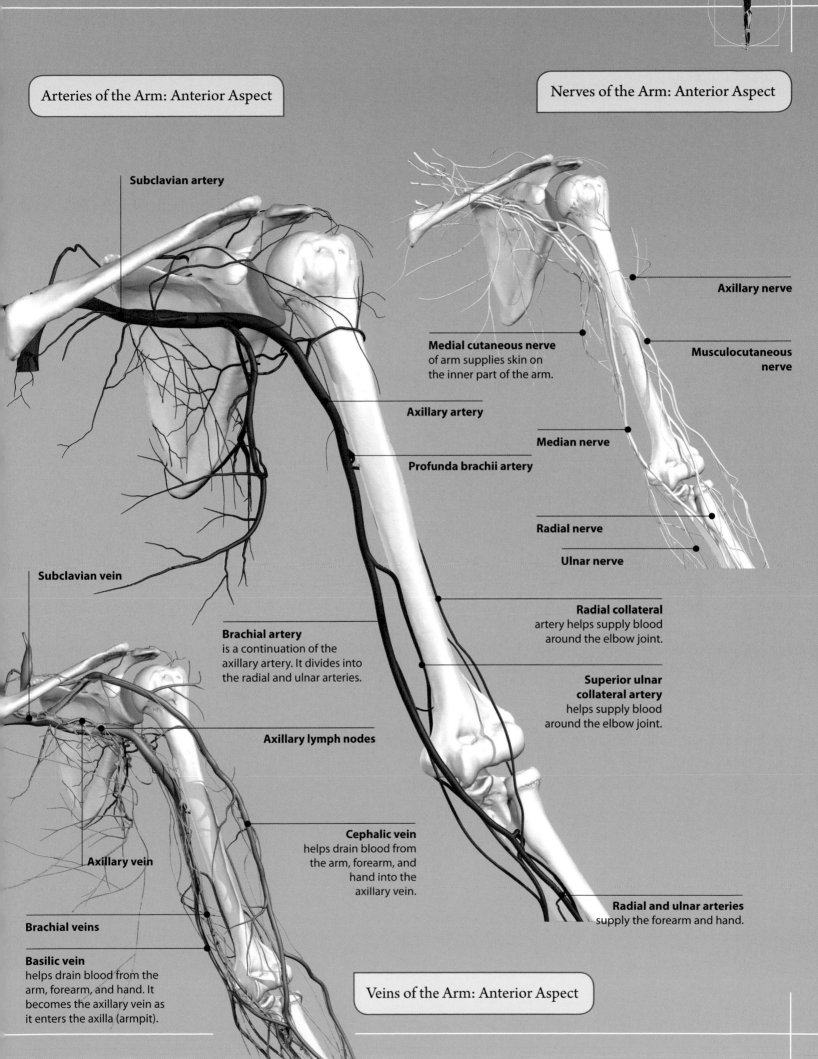

Arteries of the Arm: Anterior Aspect

Subclavian artery

Subclavian vein

Axillary artery

Profunda brachii artery

Brachial artery
is a continuation of the
axillary artery. It divides into
the radial and ulnar arteries.

Axillary lymph nodes

Nerves of the Arm: Anterior Aspect

Axillary nerve

Medial cutaneous nerve
of arm supplies skin on
the inner part of the arm.

**Musculocutaneous
nerve**

Median nerve

Radial nerve

Ulnar nerve

Radial collateral
artery helps supply blood
around the elbow joint.

**Superior ulnar
collateral artery**
helps supply blood
around the elbow joint.

Axillary vein

Cephalic vein
helps drain blood from
the arm, forearm, and
hand into the
axillary vein.

Brachial veins

Basilic vein
helps drain blood from the
arm, forearm, and hand. It
becomes the axillary vein as
it enters the axilla (armpit).

Radial and ulnar arteries
supply the forearm and hand.

Veins of the Arm: Anterior Aspect

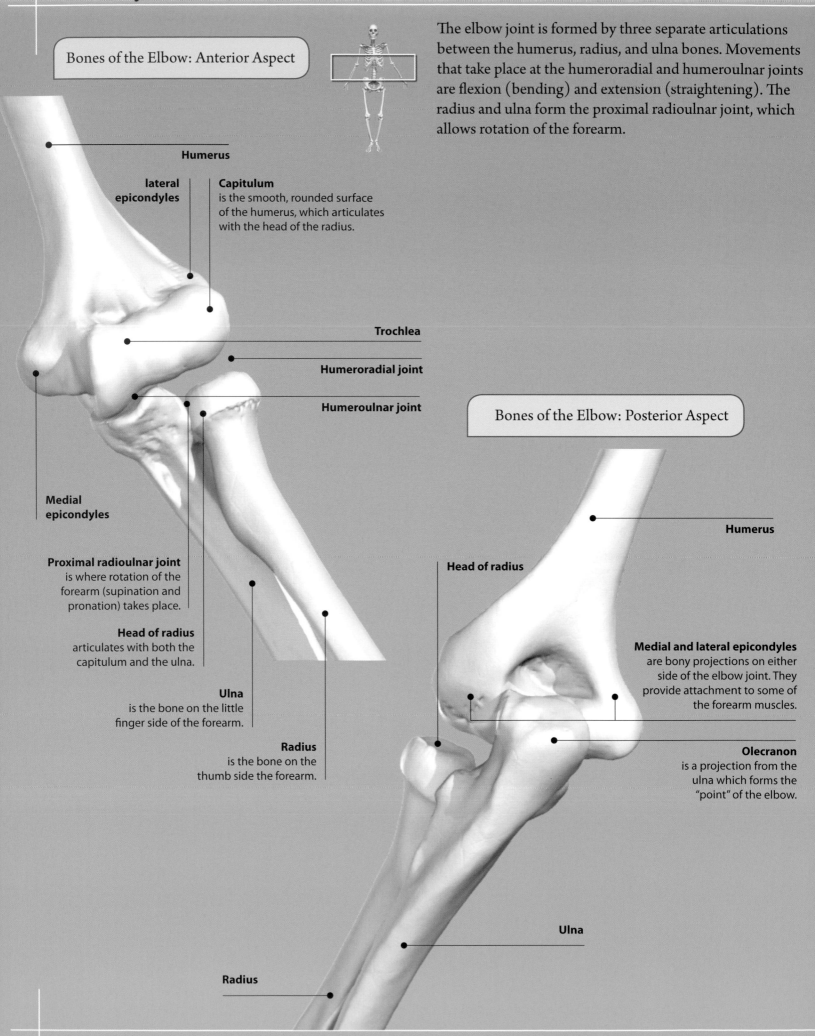

Bones of the Elbow: Anterior Aspect

The elbow joint is formed by three separate articulations between the humerus, radius, and ulna bones. Movements that take place at the humeroradial and humeroulnar joints are flexion (bending) and extension (straightening). The radius and ulna form the proximal radioulnar joint, which allows rotation of the forearm.

Humerus

lateral epicondyles

Capitulum
is the smooth, rounded surface of the humerus, which articulates with the head of the radius.

Trochlea

Humeroradial joint

Humeroulnar joint

Bones of the Elbow: Posterior Aspect

Medial epicondyles

Proximal radioulnar joint
is where rotation of the forearm (supination and pronation) takes place.

Head of radius
articulates with both the capitulum and the ulna.

Ulna
is the bone on the little finger side of the forearm.

Radius
is the bone on the thumb side the forearm.

Head of radius

Humerus

Medial and lateral epicondyles
are bony projections on either side of the elbow joint. They provide attachment to some of the forearm muscles.

Olecranon
is a projection from the ulna which forms the "point" of the elbow.

Ulna

Radius

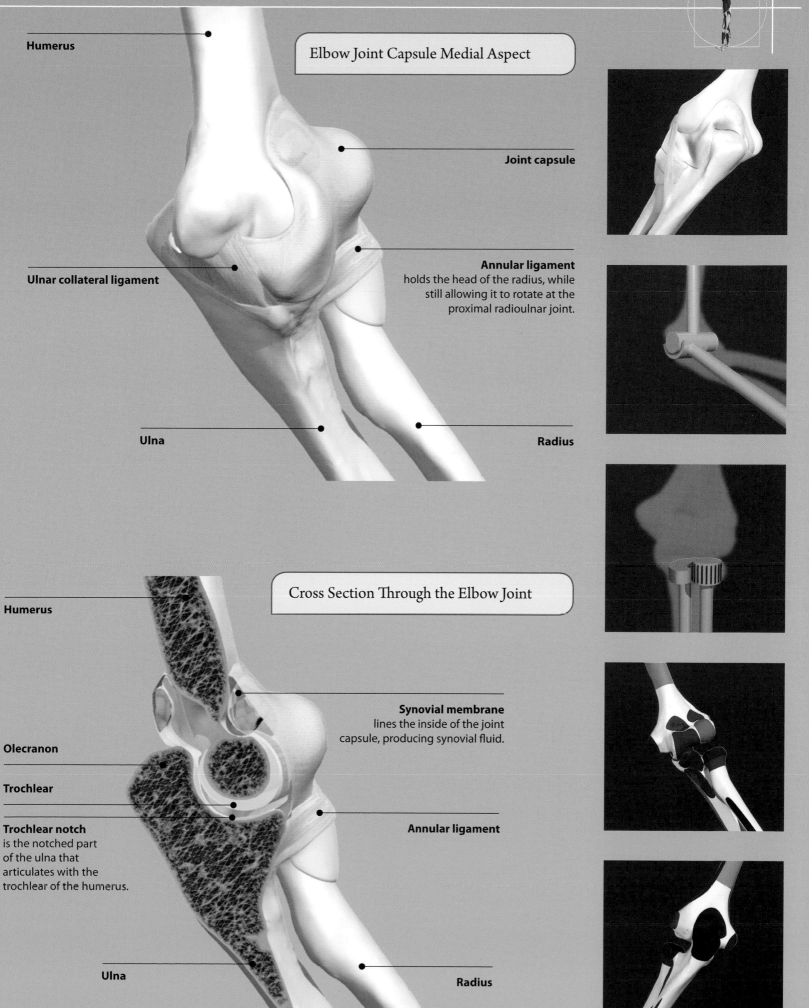

Humerus

Elbow Joint Capsule Medial Aspect

Joint capsule

Ulnar collateral ligament

Annular ligament
holds the head of the radius, while
still allowing it to rotate at the
proximal radioulnar joint.

Ulna

Radius

Cross Section Through the Elbow Joint

Humerus

Synovial membrane
lines the inside of the joint
capsule, producing synovial fluid.

Olecranon

Trochlear

Trochlear notch
is the notched part
of the ulna that
articulates with the
trochlear of the humerus.

Annular ligament

Ulna

Radius

CUBITAL FOSSA DISSECTION

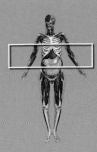

The cubital fossa is a triangular-shaped region in front of the elbow joint. The apex is formed by the brachioradialis muscle on the outside, and pronator teres muscle on the inside. The base of the triangle runs between the two bony prominences on either side of the elbow (medial and lateral epicondyles).

Many important structures pass over or through the cubital fossa as they travel between the arm and forearm.

Cephalic vein
runs along the outer part of the forearm and arm.

Deep fascia
is a tough tube of fibrous tissue. It wraps around the structures of the upper limb holding them together, and dividing the muscles into groups.

Basilic vein
runs along the inner part of the forearm and arm.

Brachial artery
divides in the cubital fossa into radial and ulnar arteries.

Elbow joint capsule
lies beneath the cubital fossa.

Median cubital vein
connects the cephalic and basilic veins. It runs over the cubital fossa, where it can be easily accessed with a needle if a blood sample is needed.

Ulnar artery

Supinator muscle
forms part of the floor of the cubital fossa. It rotates the forearm outward.

Radial artery

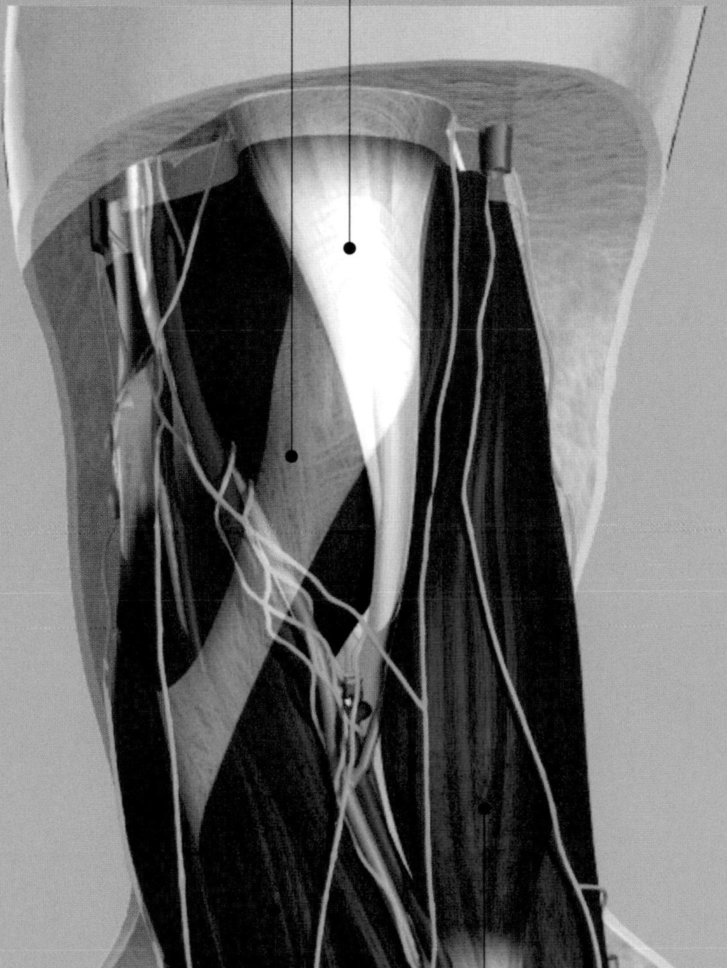

Bicipital aponeurosis
is a tough fibrous band which can be felt running over the cubital fossa. It attaches the tendon of the biceps brachii to the deep fascia.

Tendon of biceps brachii
runs through the cubital fossa to attach to the radius.

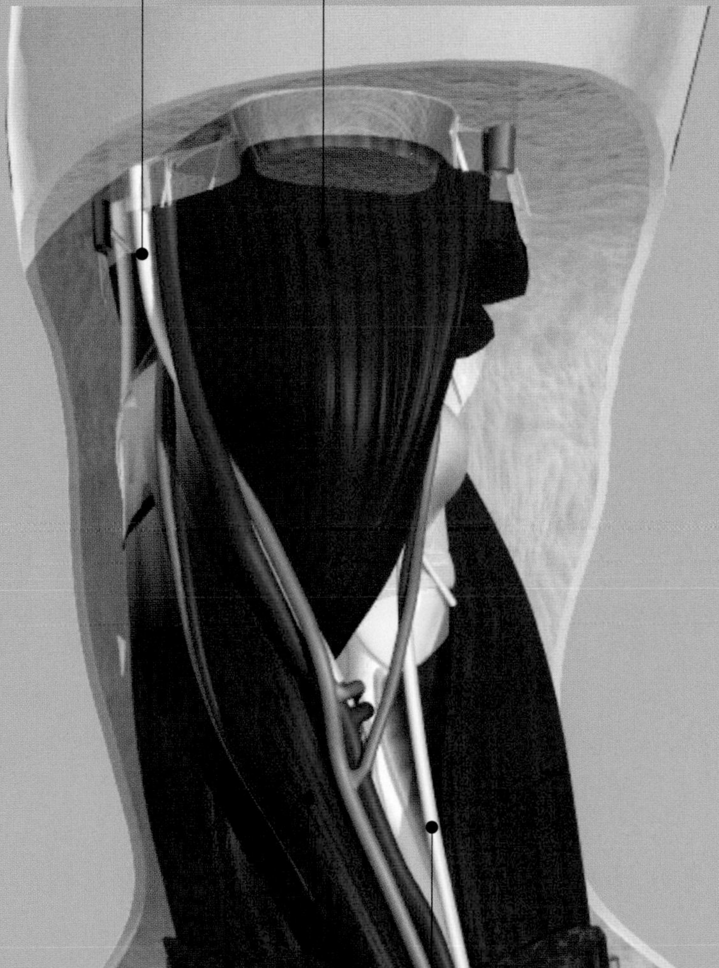

Median nerve
passes through the cubital fossa, to supply muscles that bend the fingers and thumb.

Brachialis muscle
forms the floor of the cubital fossa.

Pronator teres muscle

Brachioradialis muscle

Pronator teres muscle

Superficial radial nerve
lies behind brachioradialis and supplies sensation around the base of the thumb.

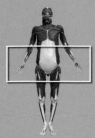

The forearm is the region of the upper limb between the elbow and the wrist. Muscles on the front of the forearm bend the fingers and wrist, and are known as flexors. Muscles on the back of the forearm straighten the fingers and wrist, and are known as extensors. These two groups can be further divided into superficial and deep muscles, depending on how close to the surface they are. The tendons from many of these muscles can be easily seen around the wrist.

Extensor retinaculum
helps guide, and hold the tendons of the extensor muscles in place.

Extensor carpi radialis brevis
assists extensor carpi radialis longus.

Extensor carpi radialis longus
helps straighten and "cock" the wrist back.

Extensor carpi ulnaris
helps straighten and "cock" the wrist back.

Extensor digiti minimi
helps straighten the little finger.

Extensor digitorum
straightens the fingers.

Brachioradialis

Abductor pollicis longus
helps lift the thumb away from the palm of the hand.

Humerus

Supinator

Ulna

Extensor indicis
helps straighten the index finger.

Extensor pollicis longus
straightens the thumb.

Radius

Extensor pollicis brevis
straightens the thumb.

Did you know?

In most people, the length of the forearm is the same as the length of the foot.

Palmaris longus
is a small muscle that helps
bend the wrist. It is absent
in some people.

Flexor carpi radialis
helps bend the wrist.

Pronator teres
helps rotate the
forearm inward.

Flexor carpi ulnaris
helps bend the wrist.

**Flexor digitorum
superficialis**
is a muscle that helps
bend the fingers.

Flexor retinaculum
helps guide and hold
the tendons of the flexor
muscles in place.

Flexor digitorum profundus
is a deep muscle that
helps bend the fingers.

Humerus

Radius

Ulna

Interosseous membrane
is a sheet of fibrous tissue
that holds the radius and
ulna together.

Pronator quadratus
helps rotate the
forearm inward.

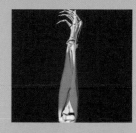

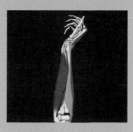

The muscles and skin of the forearm are supplied by various nerves, and blood vessels. Many of these continue through to supply structures in the hand.

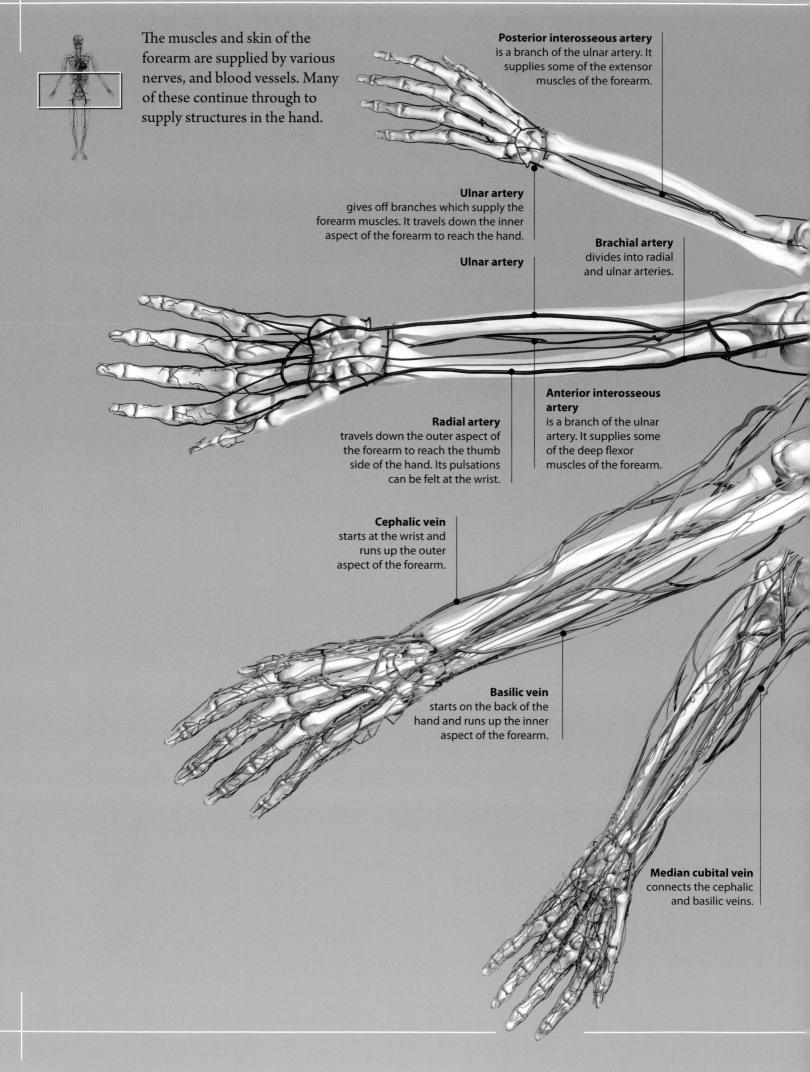

Posterior interosseous artery is a branch of the ulnar artery. It supplies some of the extensor muscles of the forearm.

Ulnar artery gives off branches which supply the forearm muscles. It travels down the inner aspect of the forearm to reach the hand.

Ulnar artery

Brachial artery divides into radial and ulnar arteries.

Radial artery travels down the outer aspect of the forearm to reach the thumb side of the hand. Its pulsations can be felt at the wrist.

Anterior interosseous artery is a branch of the ulnar artery. It supplies some of the deep flexor muscles of the forearm.

Cephalic vein starts at the wrist and runs up the outer aspect of the forearm.

Basilic vein starts on the back of the hand and runs up the inner aspect of the forearm.

Median cubital vein connects the cephalic and basilic veins.

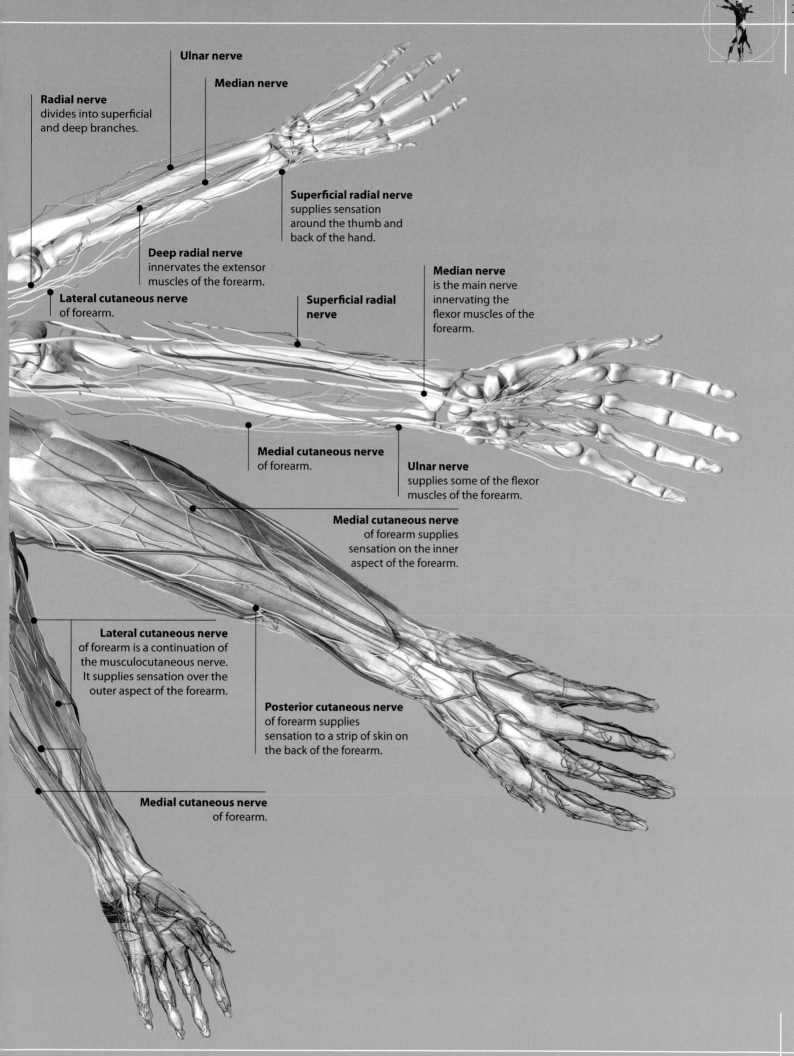

Ulnar nerve

Median nerve

Radial nerve
divides into superficial
and deep branches.

Superficial radial nerve
supplies sensation
around the thumb and
back of the hand.

Deep radial nerve
innervates the extensor
muscles of the forearm.

**Superficial radial
nerve**

Median nerve
is the main nerve
innervating the
flexor muscles of the
forearm.

Lateral cutaneous nerve
of forearm.

Medial cutaneous nerve
of forearm.

Ulnar nerve
supplies some of the flexor
muscles of the forearm.

Medial cutaneous nerve
of forearm supplies
sensation on the inner
aspect of the forearm.

Lateral cutaneous nerve
of forearm is a continuation of
the musculocutaneous nerve.
It supplies sensation over the
outer aspect of the forearm.

Posterior cutaneous nerve
of forearm supplies
sensation to a strip of skin on
the back of the forearm.

Medial cutaneous nerve
of forearm.

The hand and wrist are made up of twenty-seven bones in total. The radial and ulnar arteries supply blood to the hand and wrist, while the cephalic and basilic veins drain blood from the region.

Did you know?

"Arteriovenous anastomoses" is the name given to special channels that connect arteries directly to veins, bypassing the capillary beds. They allow blood flow to different regions of the body to be precisely controlled. In cold weather, our hands and fingers become cold and pale as blood is diverted away from them to prevent unnecessary heat loss.

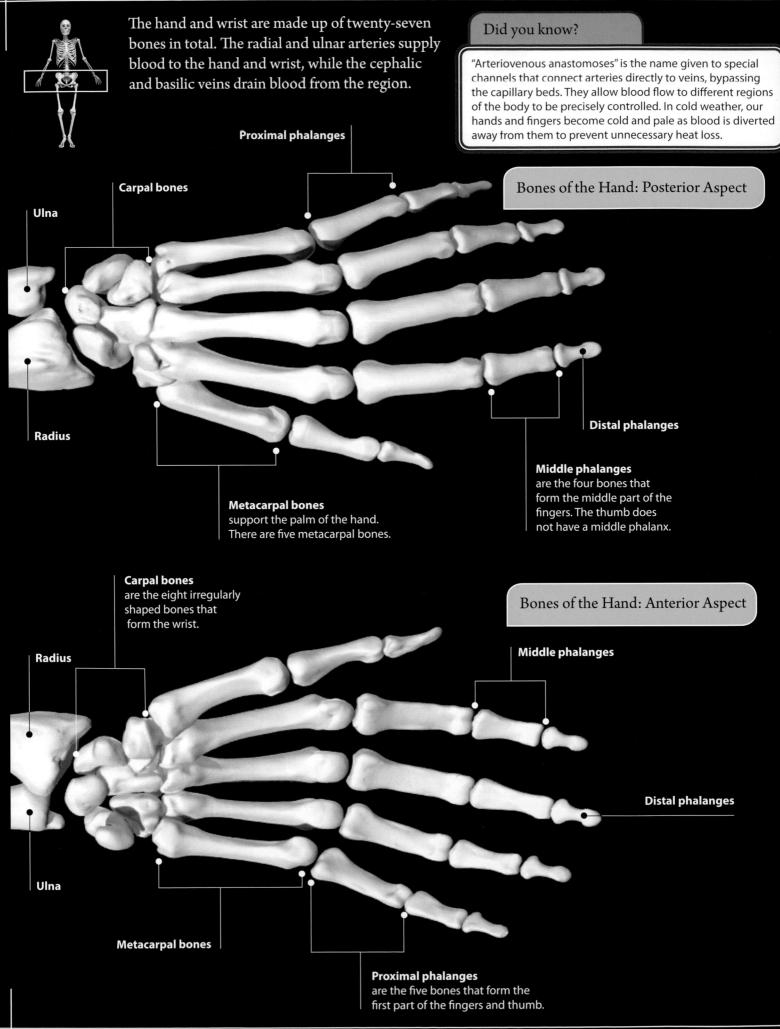

Proximal phalanges

Carpal bones

Ulna

Bones of the Hand: Posterior Aspect

Radius

Distal phalanges

Metacarpal bones
support the palm of the hand. There are five metacarpal bones.

Middle phalanges
are the four bones that form the middle part of the fingers. The thumb does not have a middle phalanx.

Carpal bones
are the eight irregularly shaped bones that form the wrist.

Bones of the Hand: Anterior Aspect

Radius

Middle phalanges

Distal phalanges

Ulna

Metacarpal bones

Proximal phalanges
are the five bones that form the first part of the fingers and thumb.

Arteries and Veins of the Hand

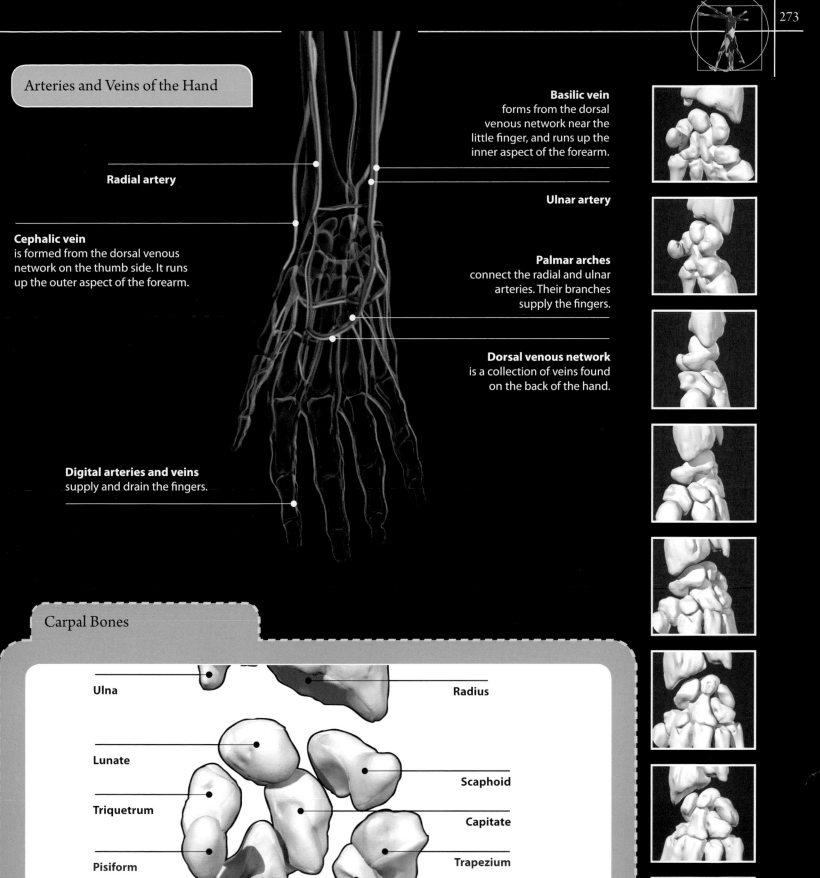

Radial artery

Basilic vein
forms from the dorsal venous network near the little finger, and runs up the inner aspect of the forearm.

Ulnar artery

Cephalic vein
is formed from the dorsal venous network on the thumb side. It runs up the outer aspect of the forearm.

Palmar arches
connect the radial and ulnar arteries. Their branches supply the fingers.

Dorsal venous network
is a collection of veins found on the back of the hand.

Digital arteries and veins
supply and drain the fingers.

Carpal Bones

Ulna

Radius

Lunate

Scaphoid

Triquetrum

Capitate

Pisiform

Trapezium

Hamate

Trapezoid

The hand contains a number of small muscles. They assist the muscles of the forearm in bending and straightening the fingers and thumb. They also allow the fingers to be spread apart (abduction) and brought back together (adduction).

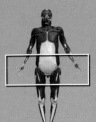

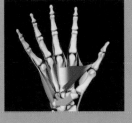

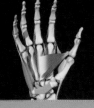

Extensor Muscles

Interossei
are small muscles found between the metacarpals of the fingers and thumb. They spread the fingers (abduction) and move them back together (adduction).

Metacarpals

Tendons of extensor digitorum
attach to the extensor hoods and straighten the fingers.

Extensor hoods
are complex fibrous structures covering the back of the fingers. Muscles attached to them help straighten the fingers.

Synovial sheaths

Did you know?

If you move your thumb out to the side, a triangular hollow can be seen to appear at the base of the thumb, formed from the extensor tendons of the thumb. It is called the "anatomical snuffbox" as it was used to hold snuff (powdered tobacco) prior to being inhaled through the nose.

Extensor retinaculum
holds down and guides the tendons of the extensor muscles of the forearm.

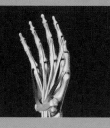
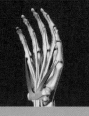

Flexor Muscles

Tendons of flexor digitorum superficialis and profundus
attach to the middle and distal phalanges of the fingers to bend them.

Lumbricals
are four small muscles that are attached between the tendons of flexor digitorum profundus and the extensor hoods. They help straighten the fingers.

Interossei

Adductor pollicis
moves the thumb towards the palm of the hand.

Synovial sheaths
are tubular structures which surround and lubricate the long tendons of some of the forearm muscles.

Flexor pollicis longus
helps bend the thumb across the palm of the hand.

Opponens pollicis
moves the thumb to touch the tip of the fingers. This movement is called opposition, and helps us carry out a wide range of tasks.

Opponens digiti minimi
helps move the little finger so that it can touch the thumb.

Flexor retinaculum
holds down the long flexor tendons of the forearm.

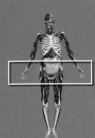

The carpal tunnel is a narrow passageway at the wrist. It is formed by the flexor retinaculum and carpal (wrist) bones. It contains nine flexor tendons and the median nerve, as they pass from the forearm into the hand.

Thenar muscles
are the group of small hand muscles at the base of the thumb.

Flexor retinaculum
is a band of tough fibrous tissue attached to the carpal bones. It forms the roof of the carpal tunnel.

Median nerve
passes through the carpal tunnel to reach the hand.

Synovial sheaths
surround and lubricate the long tendons of the forearm muscles.

Median nerve
branches supply sensation to the palm of the hand, thumb, index, and middle fingers. They also innervate the thenar muscles.

Ulnar artery and nerve
lie outside of the carpal tunnel.

Flexor retinaculum

Tendons of flexor digitorum superficialis
and profundus pass through the carpal tunnel.

Tendon of flexor pollicis longus
passes through the carpal tunnel.

Tendon of flexor pollicis longus

Wrist bones
form the floor and side walls of the carpal tunnel.

Trapezium and scaphoid
are wrist bones that provide attachment for the flexor retinaculum on the thumb side of the wrist.

Tendons of flexor digitorum superficialis and profundus

Hamate and pisiform
are wrist bones that provide attachment for the flexor retinaculum on the little finger side of the wrist.

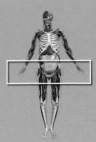

Nails are plates of dead, hardened epidermal (skin) cells that protect the ends of the fingers and toes (digits).

Fascia is the tough fibrous connective tissue that supports and holds structures in the body together. In the limbs, it separates muscles into groups that have similar actions. Fascia surrounds the muscles and allows them to move easily past each other. In certain parts of the body, the fascia is thickened to form distinct structures, such as the palmar aponeurosis.

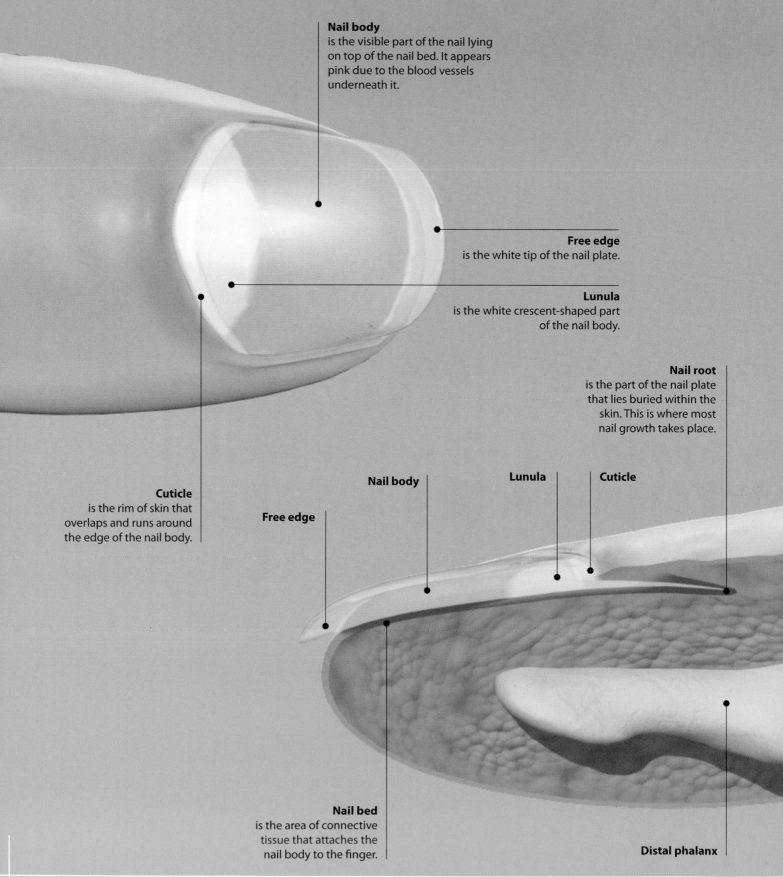

Nail body
is the visible part of the nail lying on top of the nail bed. It appears pink due to the blood vessels underneath it.

Free edge
is the white tip of the nail plate.

Lunula
is the white crescent-shaped part of the nail body.

Nail root
is the part of the nail plate that lies buried within the skin. This is where most nail growth takes place.

Cuticle
is the rim of skin that overlaps and runs around the edge of the nail body.

Free edge

Nail body

Lunula

Cuticle

Nail bed
is the area of connective tissue that attaches the nail body to the finger.

Distal phalanx

Palmar Aponeurosis

Palmar aponeurosis
is a triangular-shaped
thickening of fascia in the
palm of the hand.

Palmaris brevis
muscle attaches to the
palmar aponeurosis. It
helps improve the grip on
smooth objects.

Tendon of palmaris longus
muscle attaches into the
palmar aponeurosis.

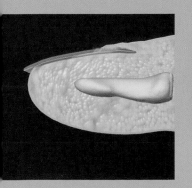
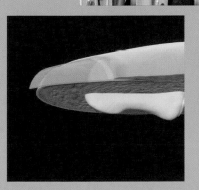
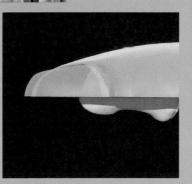
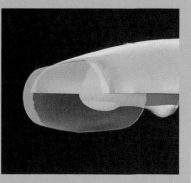

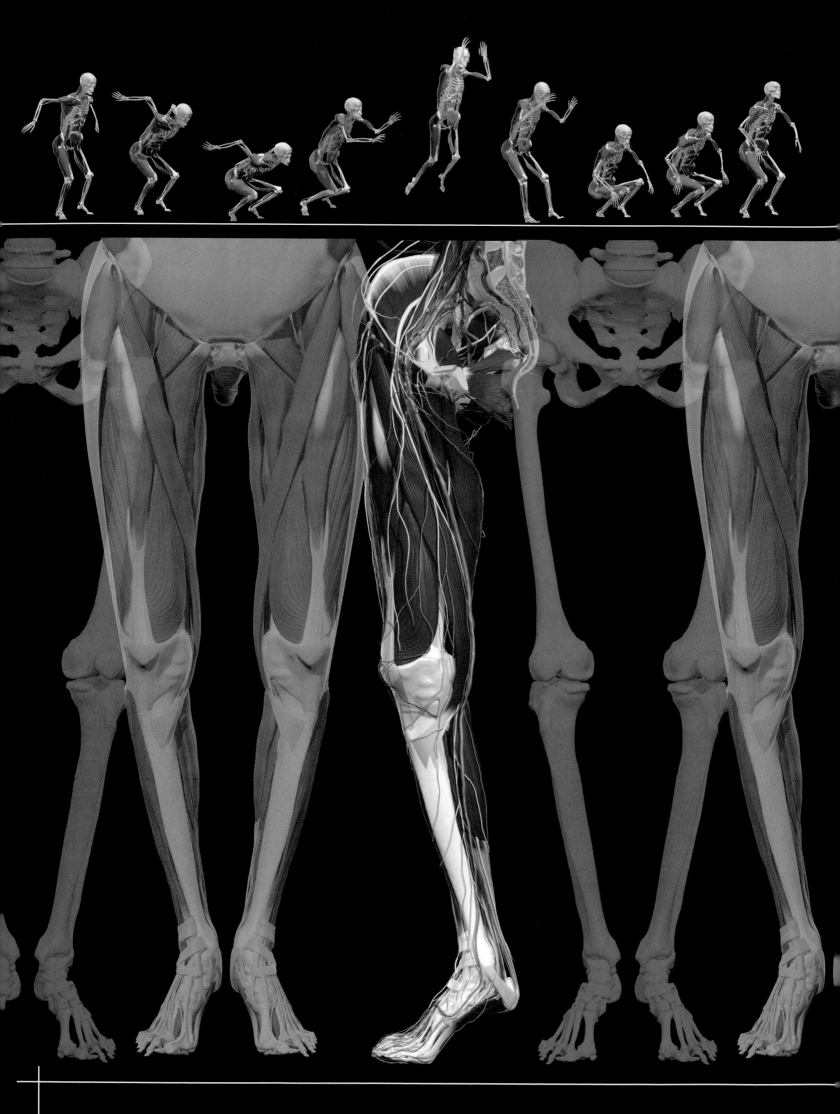

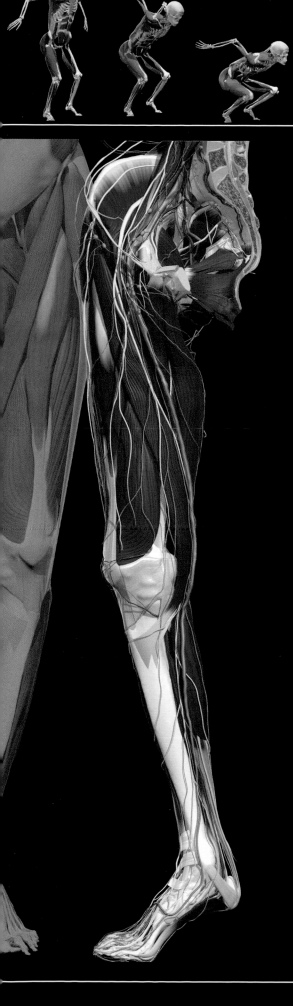

The lower limbs are attached to the trunk by strong joints between the hip bones and sacrum. The lower limb is made up of the femur (thigh bone), tibia (shin bone), fibula, and various smaller bones of the ankle and foot. It can be divided into five regions; the hip, thigh, knee, leg, and foot.

The lower limbs support the weight of the body and allow us to walk, run, and jump. This requires stable but mobile joints, along with strong muscles acting across them.

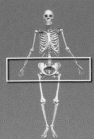

The hip joint is a ball and socket joint. It is formed as the round head of the femur sits within the cup-shaped acetabulum of the hip bone. It is a strong but mobile joint, and can be moved forward, backward, and sideways, as well as allowing some rotation. Movements are limited by various ligaments, which strengthen and stabilize the joint.

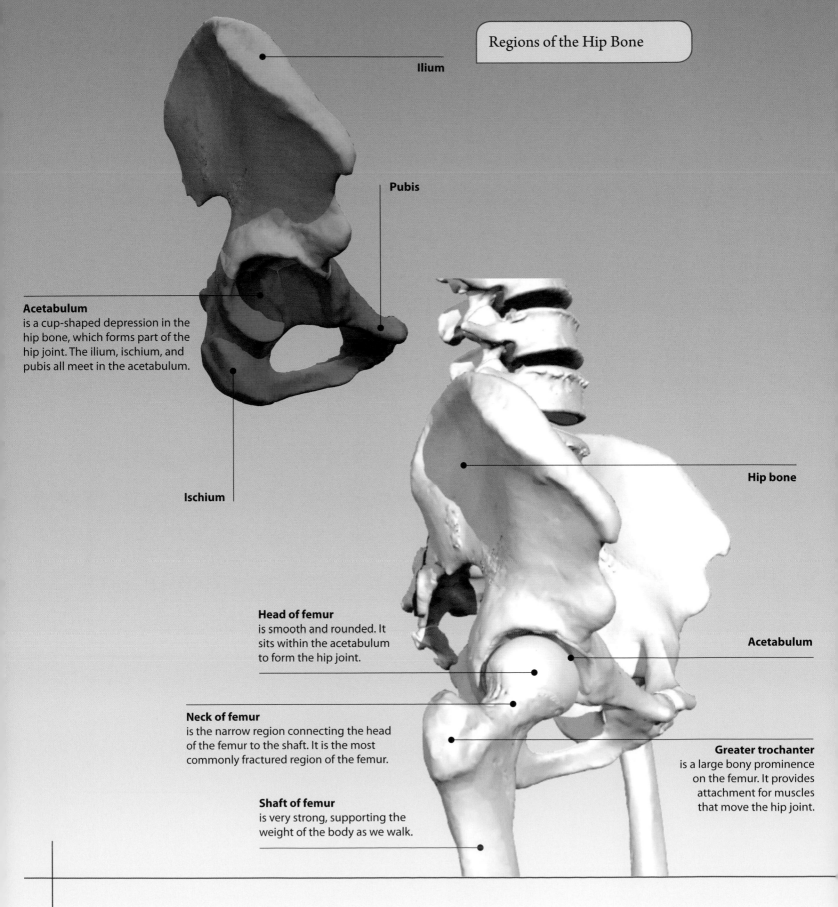

Regions of the Hip Bone

Ilium

Pubis

Acetabulum
is a cup-shaped depression in the hip bone, which forms part of the hip joint. The ilium, ischium, and pubis all meet in the acetabulum.

Ischium

Hip bone

Head of femur
is smooth and rounded. It sits within the acetabulum to form the hip joint.

Acetabulum

Neck of femur
is the narrow region connecting the head of the femur to the shaft. It is the most commonly fractured region of the femur.

Greater trochanter
is a large bony prominence on the femur. It provides attachment for muscles that move the hip joint.

Shaft of femur
is very strong, supporting the weight of the body as we walk.

Ilium

Hip Joint

Ischiofemoral ligament reinforces the hip joint at the back.

Iliofemoral ligament limits backward movements of the hip joint.

Greater trochanter

Femur, or thigh bone

Pubis

Psoas bursa is a fluid-filled sac of joint fluid. It limits friction between the hip joint and the overlying muscles.

Pubofemoral ligament reinforces the lower part of the hip joint.

Acetabulum

Acetabular labrum is made of fibrocartilage and runs around the rim of the acetabulum, increasing its depth.

Articular cartilage is horseshoe-shaped and provides a smooth joint surface within the acetabulum.

Ischiofemoral ligament

Ischium

Ilium

Iliofemoral ligament

Acetabulum

Ligamentum of head of femur attaches to the head of the femur.

Pubis

Transverse acetabular ligament bridges across the lower part of the acetabulum.

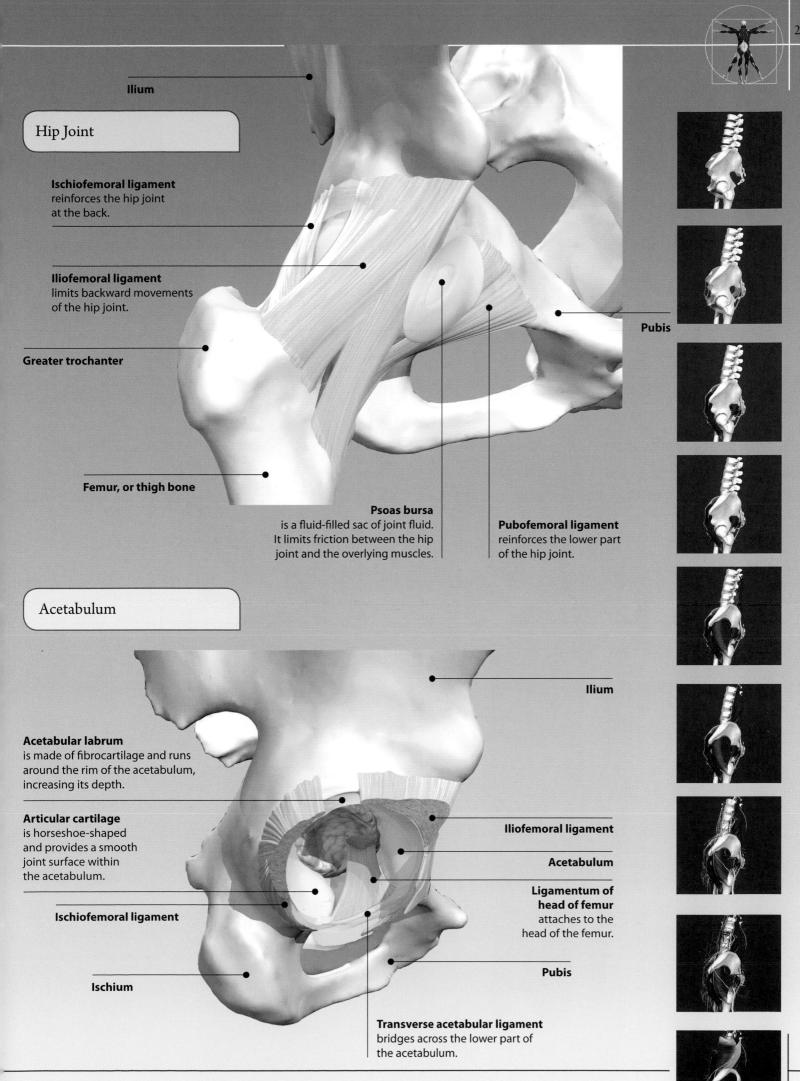

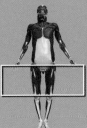

The hip is the region of the lower limb around the hip bone. The thigh is the region of the lower limb between the hip and the knee. Strong muscles in these regions move the hip and knee joints, allowing us to walk.

    The thigh can be divided into three main compartments (anterior, posterior, and medial) whose muscles all have similar actions at the hip or knee joints.

Did you know?

Vastus medialis forms the muscular bulge just above the inside of the knee when it is straightened. It is often the first muscle to shrink in size (atrophy) if the limb is injured and cannot be exercised.

Superficial Muscles of the Thigh: Anterior Aspect

Deep Muscles of the Thigh: Anterior Aspect

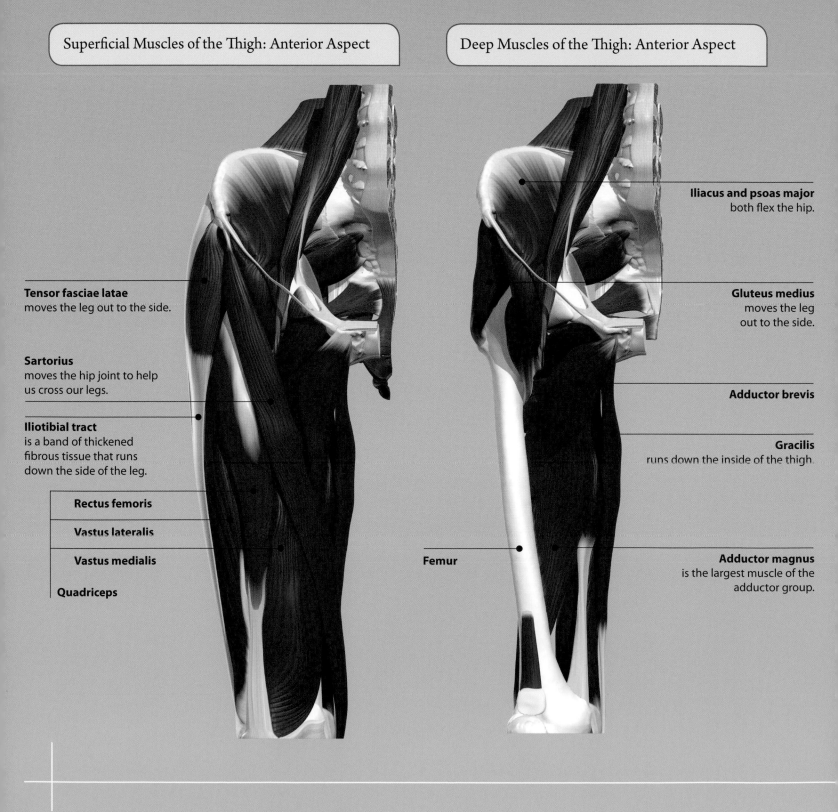

Tensor fasciae latae
moves the leg out to the side.

Sartorius
moves the hip joint to help us cross our legs.

Iliotibial tract
is a band of thickened fibrous tissue that runs down the side of the leg.

Rectus femoris

Vastus lateralis

Vastus medialis

Quadriceps

Iliacus and psoas major
both flex the hip.

Gluteus medius
moves the leg out to the side.

Adductor brevis

Gracilis
runs down the inside of the thigh.

Femur

Adductor magnus
is the largest muscle of the adductor group.

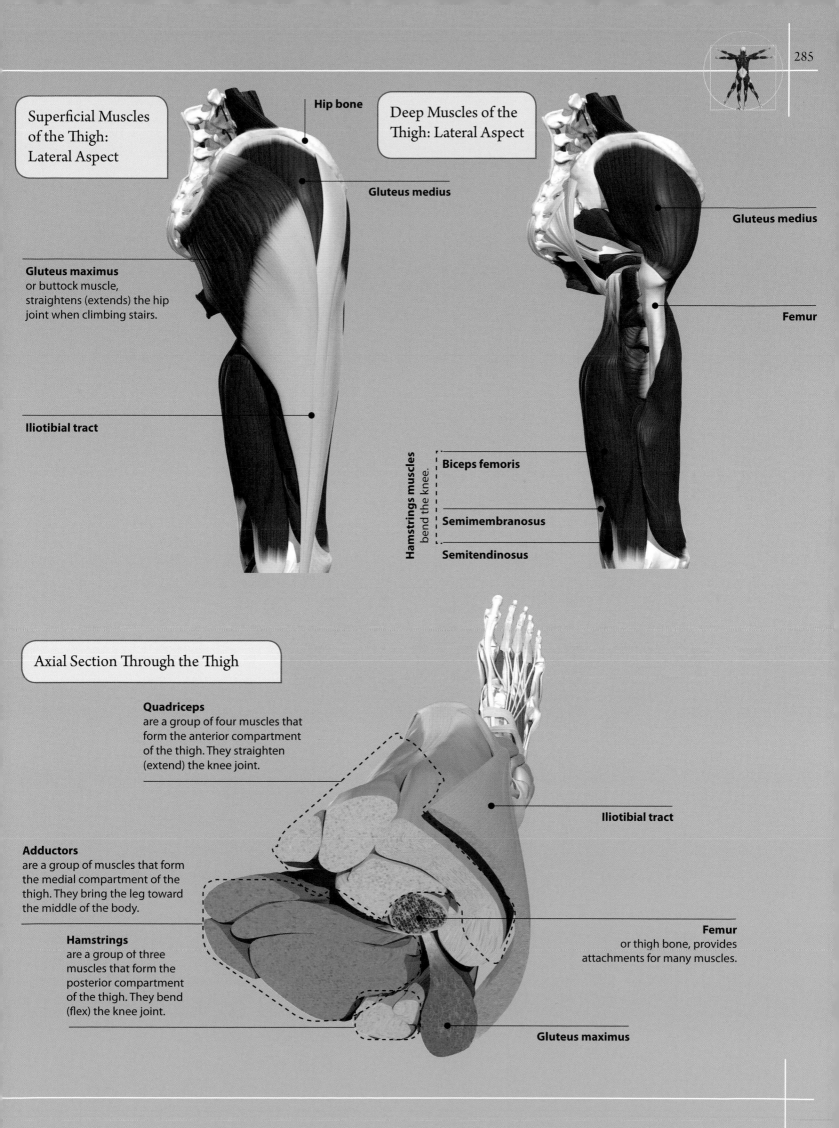

Superficial Muscles of the Thigh: Lateral Aspect

Hip bone

Gluteus medius

Gluteus maximus
or buttock muscle, straightens (extends) the hip joint when climbing stairs.

Iliotibial tract

Deep Muscles of the Thigh: Lateral Aspect

Gluteus medius

Femur

Hamstrings muscles bend the knee.

Biceps femoris

Semimembranosus

Semitendinosus

Axial Section Through the Thigh

Quadriceps
are a group of four muscles that form the anterior compartment of the thigh. They straighten (extend) the knee joint.

Iliotibial tract

Adductors
are a group of muscles that form the medial compartment of the thigh. They bring the leg toward the middle of the body.

Hamstrings
are a group of three muscles that form the posterior compartment of the thigh. They bend (flex) the knee joint.

Femur
or thigh bone, provides attachments for many muscles.

Gluteus maximus

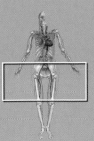

Blood is distributed to the hip and thigh by branches of the internal and external iliac vessels. Various nerves travel through the hip and thigh, supplying the muscles and skin of the lower limb.

Did you know?

If the nerve roots that supply the sciatic nerve become trapped or irritated in the lower back, the sensation of pain often radiates down the entire lower limb to the foot. This is known as sciatica.

Neurovascular Structures and Muscles of the Thigh: Anterior Aspect

Neurovascular Structures and Muscles of the Thigh: Posterior Aspect

Piriformis

External iliac artery becomes the femoral artery as it enters the thigh beneath the inguinal ligament.

Inguinal ligament

Femoral artery is the main blood supply to the lower limb.

Lateral cutaneous nerve of thigh supplies skin over the outer part of the thigh.

Long saphenous vein

Hip bone

Superior gluteal nerve and vessels exit the pelvis above piriformis to supply the upper hip region.

Gluteus minimus

Inferior gluteal nerve and muscles exit the pelvis below piriformis to supply the lower hip region.

Sciatic nerve

Posterior cutaneous nerve of thigh supplies skin at the back of the thigh.

Neurovascular Structures of the Thigh: Anterior Aspect

Hip bone

External iliac artery

Femoral artery

Lateral circumflex femoral artery
is a branch of the deep femoral artery.

Femur

Internal iliac vessels
give off the superior and inferior gluteal vessels.

Femoral vein

Deep femoral artery

Long saphenous vein
runs up the inside of the lower limb, draining blood into the femoral vein.

Popliteal artery

Neurovascular Structures of the Thigh: Posterior Aspect

Inferior gluteal vessels

Femoral artery

Deep femoral artery
is a large branch of the femoral artery.

Popliteal artery
is a continuation of the femoral artery as it moves behind the knee.

Hip bone

Superior gluteal vessels

Medial circumflex femoral artery
wraps around the femur.

Femur

Long saphenous vein

Nerves of the Hip and Thigh

Sciatic nerve
is the largest nerve in the body. It supplies the hamstring muscles before dividing into the common fibular and tibial nerves.

Obturator nerve
innervates the adductor muscles on the inner thigh.

Femoral nerve
innervates the quadriceps muscles.

Saphenous nerve
is a continuation of the femoral nerve. It supplies sensation over the inner part of the shin.

Common fibular nerve
supplies muscles that lift the toes.

Tibial nerve
supplies the calf muscles at the back of the lower leg.

KNEE JOINT

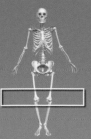

The knee joint is a complex hinge joint, formed between the femur (thigh bone), tibia (shin bone), and patella (kneecap). The main movements at the knee are flexion (bending) and extension (straightening), along with a small degree of rotation.

Bones of the Knee: Anterior Aspect

Femur

Articular cartilage
lines the parts of the bones that come in contact with each other at the joint, providing a smooth surface.

Tibial tuberosity
is an elevated area on the tibia, just below the knee. It provides attachment for the patellar ligament, allowing the knee to be straightened by the quadriceps.

Tibia

Bones of the Knee: Posterior Aspect

Femur

Femoral condyles
are the expanded lower ends of the femur that form part of the knee joint.

Tibial plateau
is the flattened upper part of the tibia, which forms part of the knee joint.

Tibia

Fibula

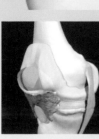
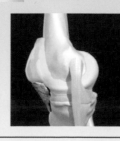
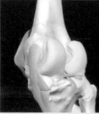

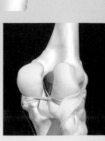

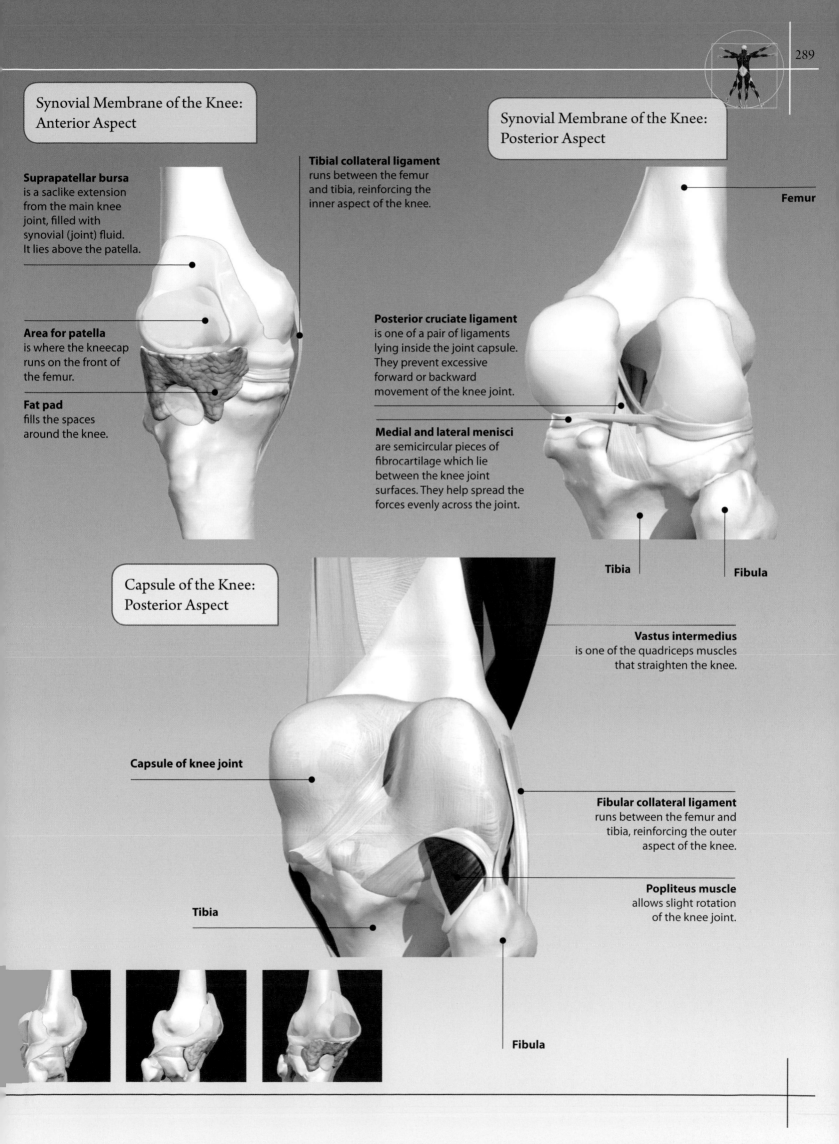

Synovial Membrane of the Knee: Anterior Aspect

Suprapatellar bursa
is a saclike extension from the main knee joint, filled with synovial (joint) fluid. It lies above the patella.

Area for patella
is where the kneecap runs on the front of the femur.

Fat pad
fills the spaces around the knee.

Tibial collateral ligament
runs between the femur and tibia, reinforcing the inner aspect of the knee.

Synovial Membrane of the Knee: Posterior Aspect

Femur

Posterior cruciate ligament
is one of a pair of ligaments lying inside the joint capsule. They prevent excessive forward or backward movement of the knee joint.

Medial and lateral menisci
are semicircular pieces of fibrocartilage which lie between the knee joint surfaces. They help spread the forces evenly across the joint.

Tibia

Fibula

Capsule of the Knee: Posterior Aspect

Capsule of knee joint

Tibia

Vastus intermedius
is one of the quadriceps muscles that straighten the knee.

Fibular collateral ligament
runs between the femur and tibia, reinforcing the outer aspect of the knee.

Popliteus muscle
allows slight rotation of the knee joint.

Fibula

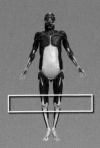

The knee is the region of the lower limb between the thigh and lower leg. Muscles crossing the knee allow movement to take place at the joint. The main movements are flexion (bending) and extension (straightening), along with a small amount of rotation.

Coronal Section of the Knee

Biceps femoris
is one of the hamstring muscles. It helps flex the knee.

Cruciate ligaments
are paired ligaments attached to the femur and tibia. They prevent excessive forward or backward movements of the knee joint.

Tibia (shin bone)

Tibial Meniscus

Patella

Tibial plateau
is the flattened upper part of the tibia.

Menisci
are curved pieces of fibrocartilage that lie between the surfaces of the knee joint.

Tibial nerve
is a branch of the sciatic nerve. It supplies the calf muscles.

Iliotibial tract
helps keep the knee locked in a straight position when standing.

Cruciate ligaments

Tendon of biceps femoris
attaches to the tibia. Biceps femoris is one of the hamstring muscles, and bends the knee.

Common fibular nerve
runs around the outside of the leg, and supplies lower leg muscles.

Sagittal Section of the Knee

Quadriceps tendon
attaches the quadriceps muscles to the patella. Using this tendon and the patellar ligament, the quadriceps muscles extend the knee.

Patella (kneecap)
lies on the front of the femur.

Anterior cruciate ligament
is attached to the front part of the tibial plateau.

Patellar ligament
attaches the patella to the tibia.

Tibia

Femur

Hamstring muscles

Tibial plateau

Popliteal vessels
pass behind the knee joint.

Tibial nerve

Popliteus
is a muscle which helps rotate the knee, and unlock it from the straightened position.

Gastrocnemius

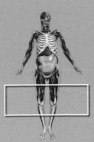

The popliteal fossa is the diamond-shaped hollow at the back of the knee. Its boundaries are formed by muscles of the back of the thigh (hamstrings) and lower leg (calf). It contains blood vessels, nerves, and lymph nodes, traveling to and from the lower leg.

Did you know?

The sural nerve can be used as a nerve graft to repair damaged nerves elsewhere in the body. It is often chosen because it does not supply any muscles, and the area of skin it supplies is relatively small on the back of the leg.

Popliteal Fossa

Deep fascia
is the fibrous tissue that encloses the muscles of the lower limb.

Popliteal fossa
is best seen when the knee is slightly bent.

Long saphenous vein
travels up the inside of the lower limb.

Sural nerve
supplies skin on the back of the calf.

Short saphenous vein
travels up the back of the calf. It helps drain blood from the foot and lower leg. It pierces the deep fascia of the popliteal fossa to drain into the popliteal vein.

Popliteal Fossa Dissection

Biceps femoris
is a hamstring muscle that
forms the upper outer
boundary of the
popliteal fossa.

**Semimembranosus and
semitendinosus**
are hamstring muscles.
They form the upper inner
boundary of the popliteal fossa.

Gastrocnemius
is one of the calf muscles.
Its two heads form the
lower boundaries of the
popliteal fossa.

Adductor magnus muscle
helps flex the knee joint.

Adductor hiatus
is an opening in the adductor
magnus. The femoral vessels pass
from the thigh into the popliteal
fossa through this opening,
becoming the popliteal vessels.

Popliteal vessels
supply the knee and lower leg.

Sciatic nerve
innervates the hamstrings
and lower leg muscles.

Gastrocnemius

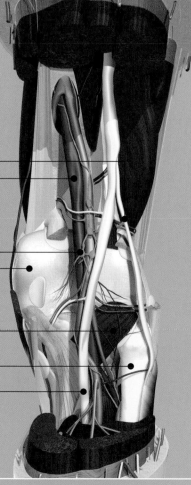

Common fibular nerve
winds around the neck of the
fibula to reach the lower leg.

Popliteal vessels

Popliteal lymph nodes
drain tissue fluid from the
back of the lower leg.

Knee joint capsule
forms part of the floor
of the popliteal fossa.

Popliteus

Fibula

Tibial nerve
travels through the popliteal fossa
to reach the calf muscles.

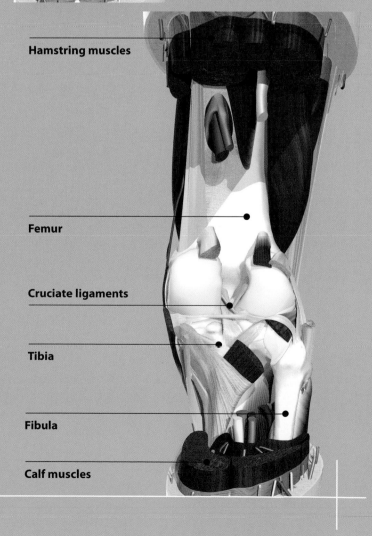

Hamstring muscles

Femur

Cruciate ligaments

Tibia

Fibula

Calf muscles

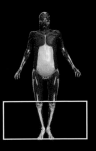

The leg is the region of the lower limb between the ankle and the knee, formed by the tibia (shin bone) and fibula.

The muscles of the leg are divided into three compartments. Their long tendons attach them to the foot bones, causing movement at the ankle and toe joints.

Cross–section through Calf to Show Compartments

Posterior compartment leg muscles are split into superficial and deep groups. They raise the heel and bend the toes, allowing us to push off the floor when we walk.

Tendons

Anterior compartment leg muscles raise the foot off the floor and straighten (extend) the toes.

Lateral compartment leg muscles turn the sole of the foot to face inwards.

Knee

Leg

Tibia (shin bone)

Ankle

Foot

Femur

Tibia

Fibula is found on the outer part of the leg. It provides attachment for muscles of the leg.

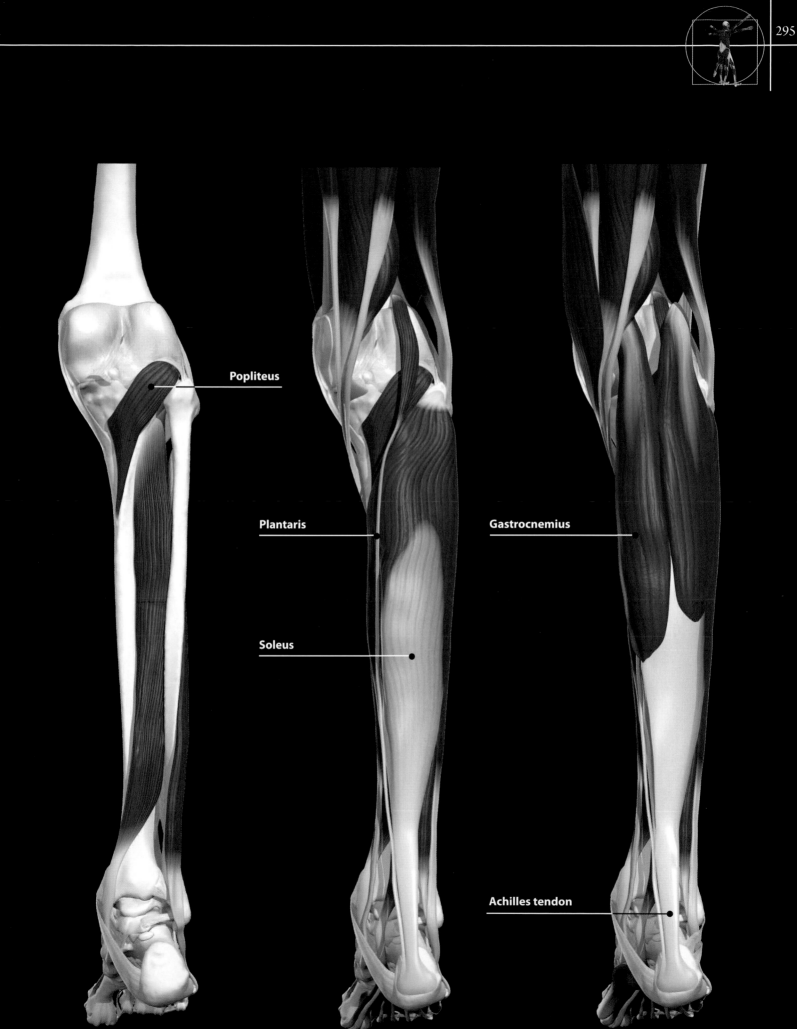

Popliteus

Plantaris

Gastrocnemius

Soleus

Achilles tendon

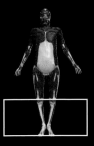

Branches of the popliteal artery supply the compartments of the leg with blood, while branches of the sciatic nerve innvervate the muscles. Blood drains from the leg via both deep and superficial veins.

Anterior

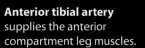

Anterior tibial artery
supplies the anterior
compartment leg muscles.

Long saphenous vein
runs up the inside
of the leg.

Dorsal venous arch

Tibia

Long saphenous vein

**Anterior
compartment
leg muscles**

Dorsal venous arch
is found on the back of the
foot. It drains into the long
and short saphenous veins.

Superficial fibular nerve
innervates the
lateral compartment.

Deep fibular nerve
innervates the
anterior compartment.

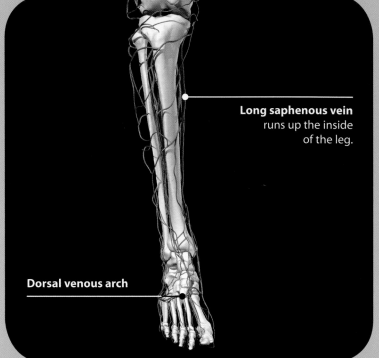

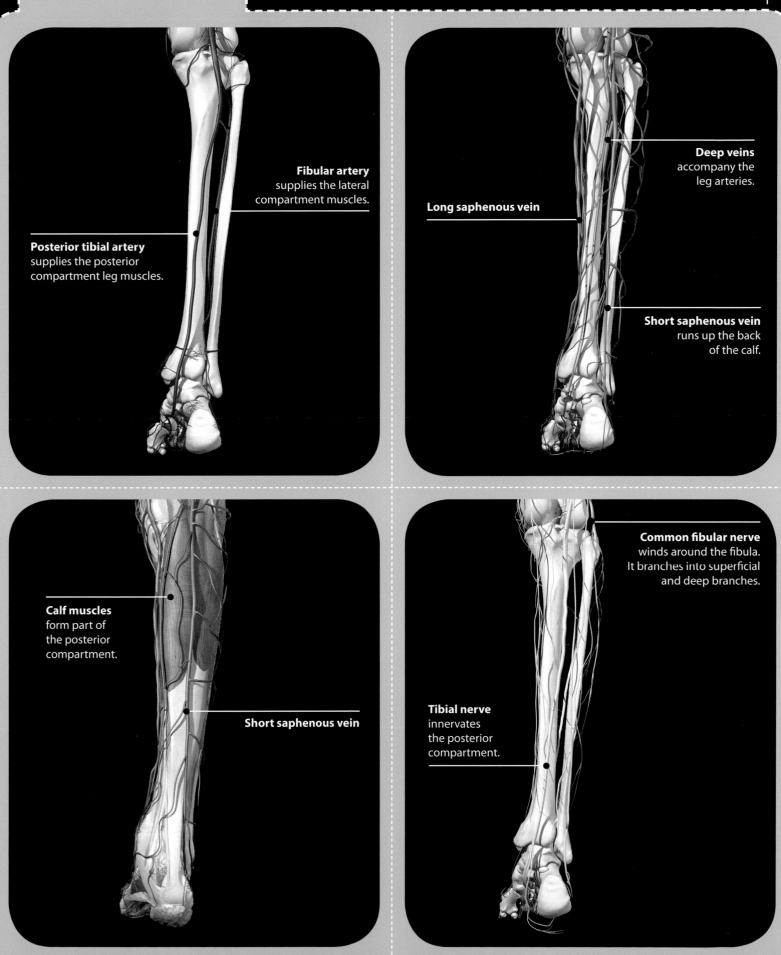

Fibular artery supplies the lateral compartment muscles.

Posterior tibial artery supplies the posterior compartment leg muscles.

Deep veins accompany the leg arteries.

Long saphenous vein

Short saphenous vein runs up the back of the calf.

Calf muscles form part of the posterior compartment.

Short saphenous vein

Common fibular nerve winds around the fibula. It branches into superficial and deep branches.

Tibial nerve innervates the posterior compartment.

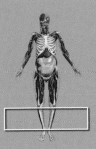

The calf is the back part of the leg, formed by the posterior compartment muscles. These muscles bend (flex) the toes, as well as raising the heel from the floor. They are divided into superficial and deep groups by a layer of fascia.

Did you know?

The ankle jerk is an involuntary reflex contraction of the calf muscles when the Achilles tendon is tapped with a tendon hammer. The ankle jerk can be used by doctors to assess if the nerves supplying the calf muscles are intact.

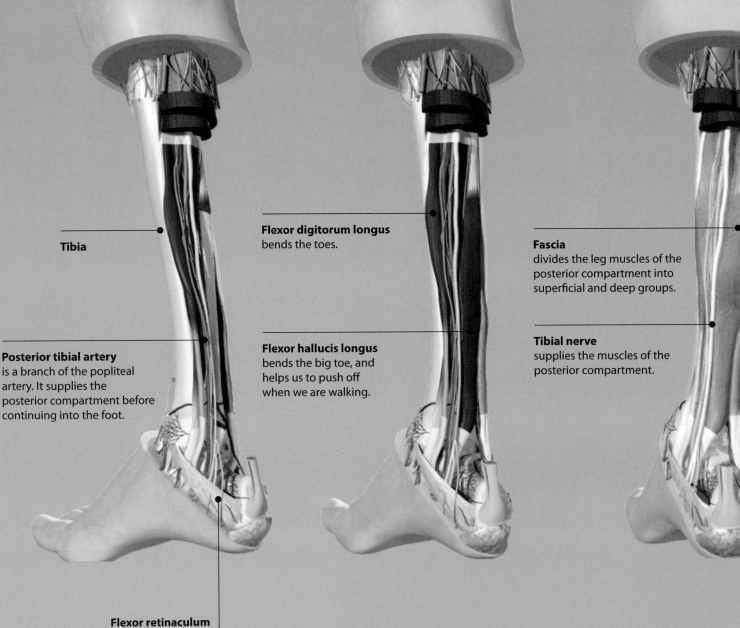

Tibia

Posterior tibial artery
is a branch of the popliteal artery. It supplies the posterior compartment before continuing into the foot.

Flexor retinaculum
helps guide the long tendons of the posterior compartment muscles as they pass behind the ankle.

Flexor digitorum longus
bends the toes.

Flexor hallucis longus
bends the big toe, and helps us to push off when we are walking.

Fascia
divides the leg muscles of the posterior compartment into superficial and deep groups.

Tibial nerve
supplies the muscles of the posterior compartment.

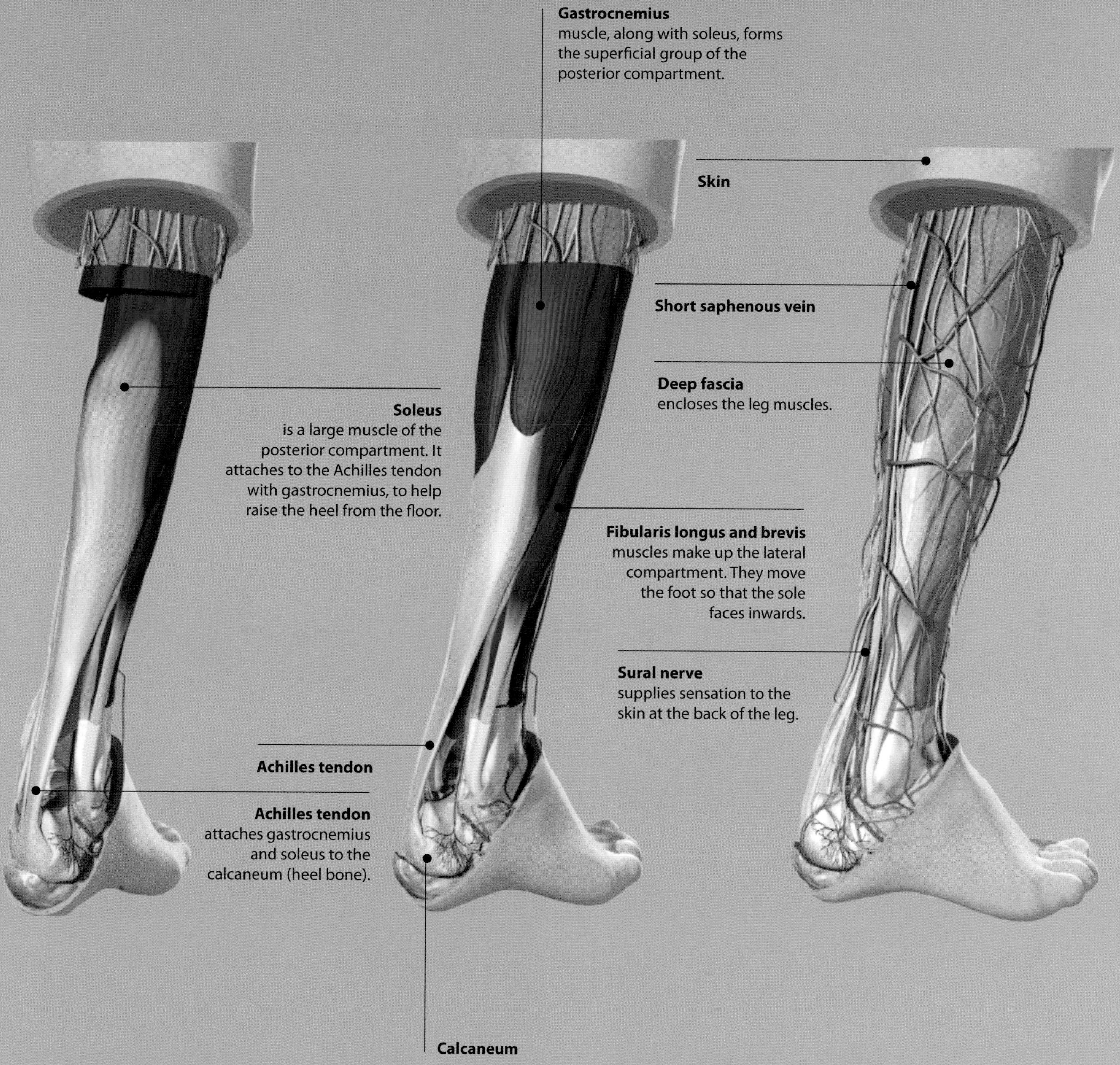

Gastrocnemius
muscle, along with soleus, forms the superficial group of the posterior compartment.

Skin

Short saphenous vein

Deep fascia
encloses the leg muscles.

Soleus
is a large muscle of the posterior compartment. It attaches to the Achilles tendon with gastrocnemius, to help raise the heel from the floor.

Fibularis longus and brevis
muscles make up the lateral compartment. They move the foot so that the sole faces inwards.

Sural nerve
supplies sensation to the skin at the back of the leg.

Achilles tendon

Achilles tendon
attaches gastrocnemius and soleus to the calcaneum (heel bone).

Calcaneum

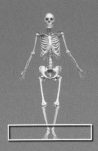

The ankle and foot make up the bottom part of the lower limb. The ankle is a hinge joint formed by the tibia, fibula, and talus, and reinforced by numerous ligaments.

The foot contains twenty-six bones in total. These bones form arches, which help spread some of the forces as we walk.

Bones of the Foot: Dorsal Aspect

Bones of the Foot: Plantar Aspect

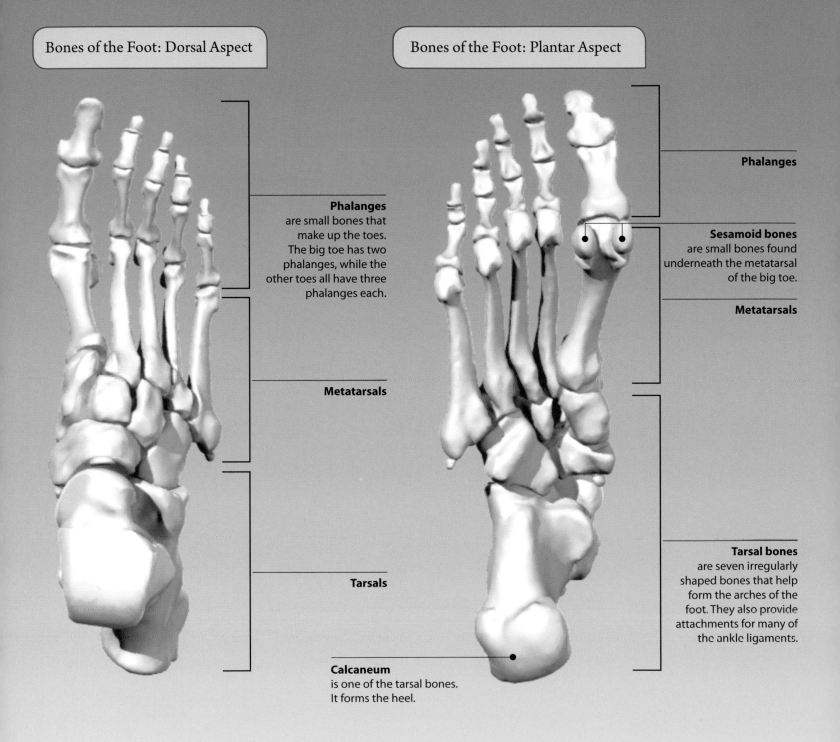

Phalanges
are small bones that make up the toes. The big toe has two phalanges, while the other toes all have three phalanges each.

Metatarsals

Tarsals

Calcaneum
is one of the tarsal bones. It forms the heel.

Phalanges

Sesamoid bones
are small bones found underneath the metatarsal of the big toe.

Metatarsals

Tarsal bones
are seven irregularly shaped bones that help form the arches of the foot. They also provide attachments for many of the ankle ligaments.

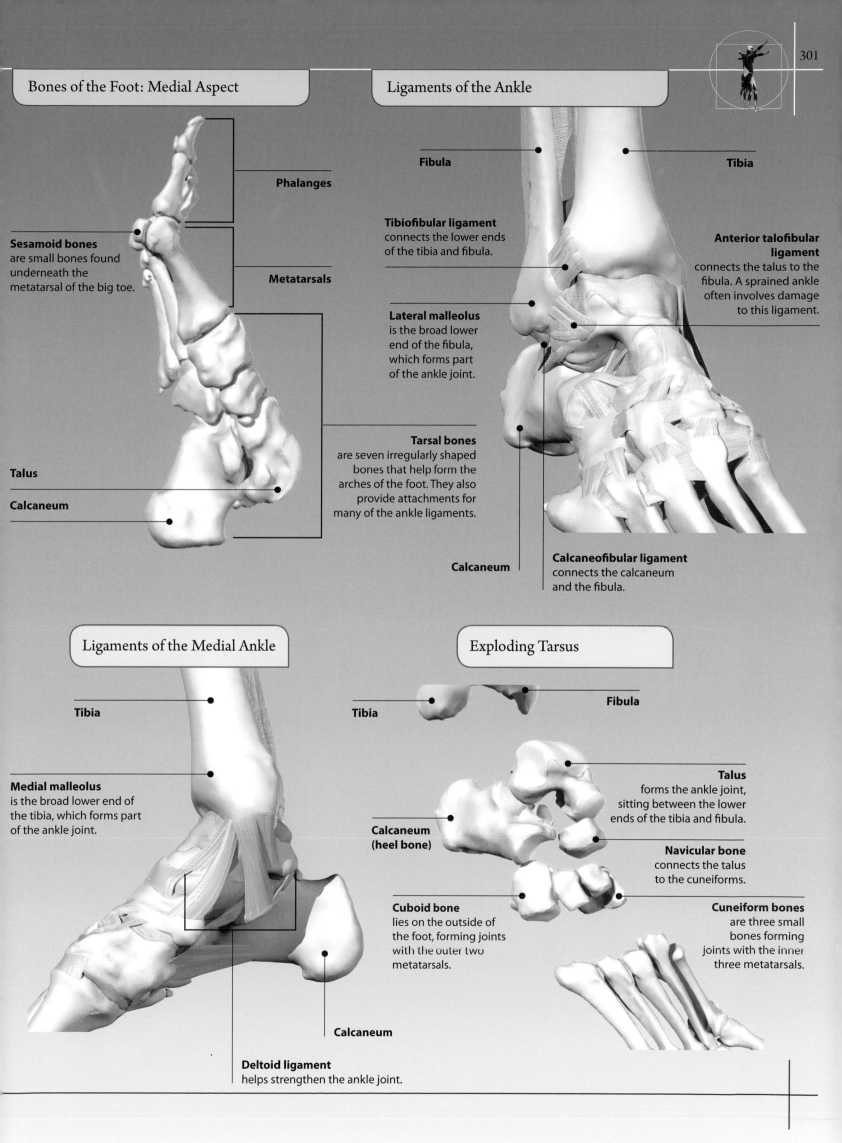

Bones of the Foot: Medial Aspect

Phalanges

Sesamoid bones
are small bones found
underneath the
metatarsal of the big toe.

Metatarsals

Talus

Calcaneum

Tarsal bones
are seven irregularly shaped
bones that help form the
arches of the foot. They also
provide attachments for
many of the ankle ligaments.

Ligaments of the Ankle

Fibula

Tibia

Tibiofibular ligament
connects the lower ends
of the tibia and fibula.

**Anterior talofibular
ligament**
connects the talus to the
fibula. A sprained ankle
often involves damage
to this ligament.

Lateral malleolus
is the broad lower
end of the fibula,
which forms part
of the ankle joint.

Calcaneum

Calcaneofibular ligament
connects the calcaneum
and the fibula.

Ligaments of the Medial Ankle

Tibia

Medial malleolus
is the broad lower end of
the tibia, which forms part
of the ankle joint.

Calcaneum

Deltoid ligament
helps strengthen the ankle joint.

Exploding Tarsus

Tibia

Fibula

Talus
forms the ankle joint,
sitting between the lower
ends of the tibia and fibula.

**Calcaneum
(heel bone)**

Navicular bone
connects the talus
to the cuneiforms.

Cuboid bone
lies on the outside of
the foot, forming joints
with the outer two
metatarsals.

Cuneiform bones
are three small
bones forming
joints with the inner
three metatarsals.

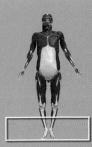

Muscles of the leg move both the ankle and the toes. They are helped in moving the toes by the muscles of the foot. These small muscles are divided into two main groups: those on the top of the foot (dorsum) and those on the sole of the foot (plantar).

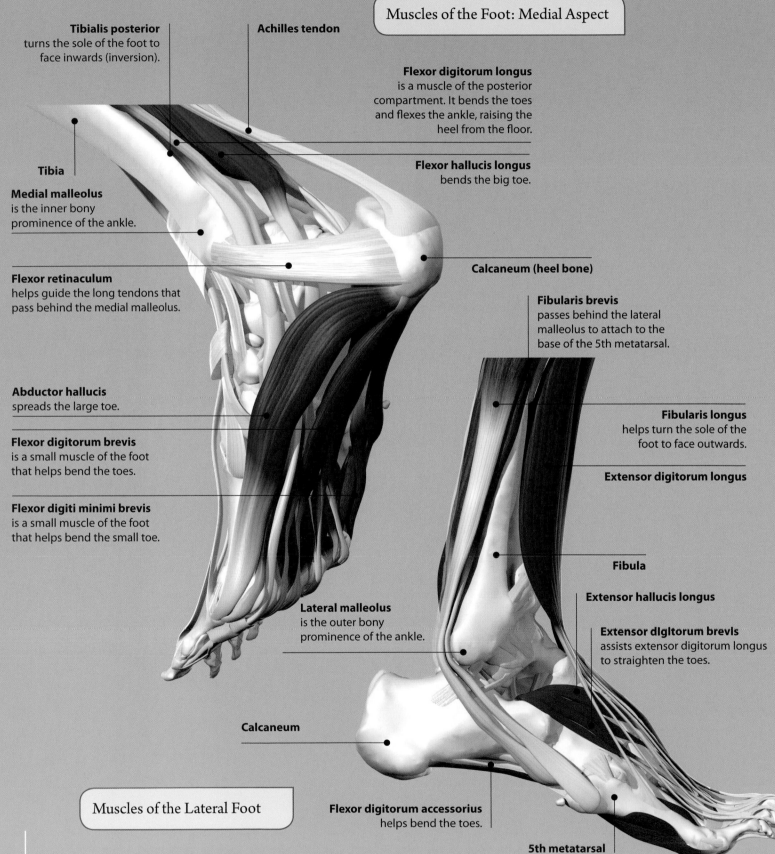

Muscles of the Foot: Medial Aspect

Tibialis posterior
turns the sole of the foot to face inwards (inversion).

Achilles tendon

Flexor digitorum longus
is a muscle of the posterior compartment. It bends the toes and flexes the ankle, raising the heel from the floor.

Tibia

Flexor hallucis longus
bends the big toe.

Medial malleolus
is the inner bony prominence of the ankle.

Calcaneum (heel bone)

Flexor retinaculum
helps guide the long tendons that pass behind the medial malleolus.

Fibularis brevis
passes behind the lateral malleolus to attach to the base of the 5th metatarsal.

Abductor hallucis
spreads the large toe.

Fibularis longus
helps turn the sole of the foot to face outwards.

Flexor digitorum brevis
is a small muscle of the foot that helps bend the toes.

Extensor digitorum longus

Flexor digiti minimi brevis
is a small muscle of the foot that helps bend the small toe.

Fibula

Extensor hallucis longus

Lateral malleolus
is the outer bony prominence of the ankle.

Extensor digitorum brevis
assists extensor digitorum longus to straighten the toes.

Calcaneum

Muscles of the Lateral Foot

Flexor digitorum accessorius
helps bend the toes.

5th metatarsal

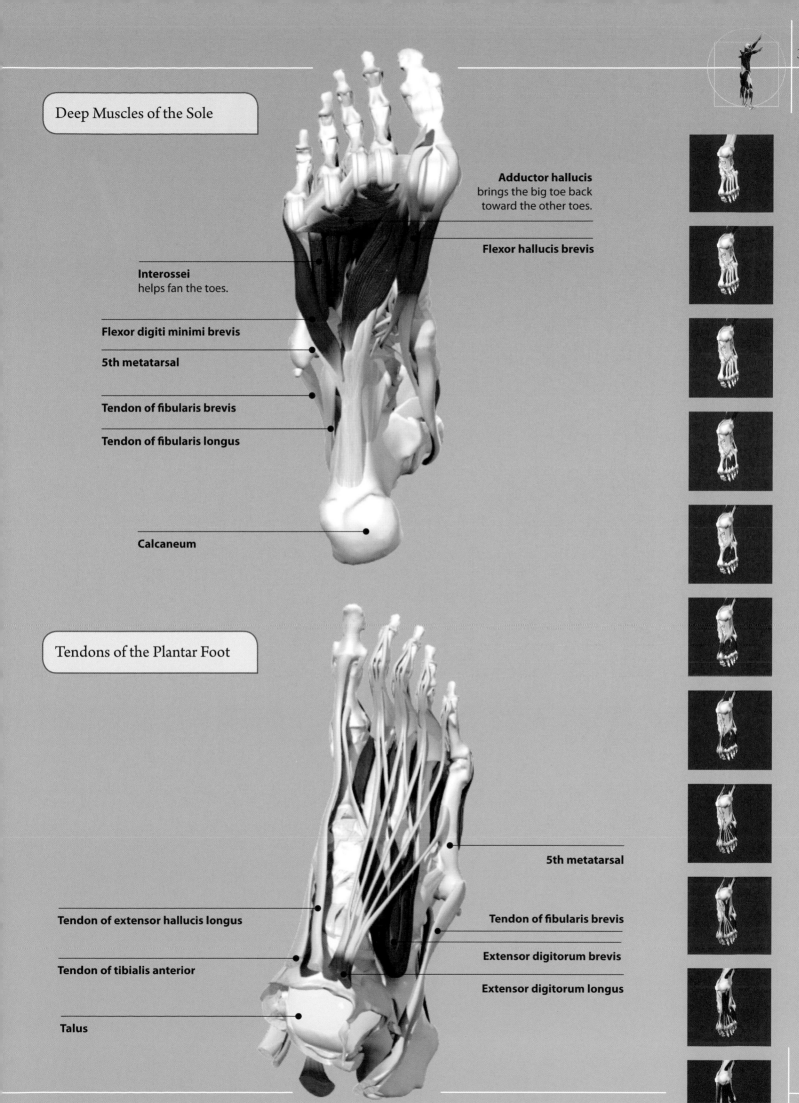

Deep Muscles of the Sole

Adductor hallucis
brings the big toe back
toward the other toes.

Flexor hallucis brevis

Interossei
helps fan the toes.

Flexor digiti minimi brevis

5th metatarsal

Tendon of fibularis brevis

Tendon of fibularis longus

Calcaneum

Tendons of the Plantar Foot

5th metatarsal

Tendon of extensor hallucis longus

Tendon of fibularis brevis

Extensor digitorum brevis

Tendon of tibialis anterior

Extensor digitorum longus

Talus

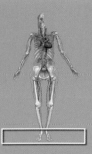

Arteries and veins travel the length of the lower limb, carrying blood to and from the foot. Various nerves supply sensation to upper and lower surfaces of the foot.

Fibular artery
supplies the lateral compartment leg muscles.

Posterior tibial artery
supplies the posterior compartment leg muscles before passing behind the medial malleolus. It branches in the foot into medial and lateral plantar arteries.

Medial malleolus
is the bony prominence on the inside of the ankle.

Anterior tibial artery
is a branch of the popliteal artery. It supplies the anterior compartment leg muscles, before continuing into the foot.

Medial plantar artery
supplies the inner part of the sole.

Lateral plantar artery
supplies the outer part of the sole.

Dorsalis pedis artery
is the continuation of the anterior tibial artery as it runs over the upper surface of the foot.

Did you know?

The pulsations of the posterior tibial artery can be felt as it passes behind the medial malleolus. The pulsations of the dorsalis pedis artery can be felt on the top of the foot as it runs alongside the tendon to the big toe.

Arcuate artery
curves over the upper surface of the foot, giving off branches to the toes.

Digital arteries
supply the toes.

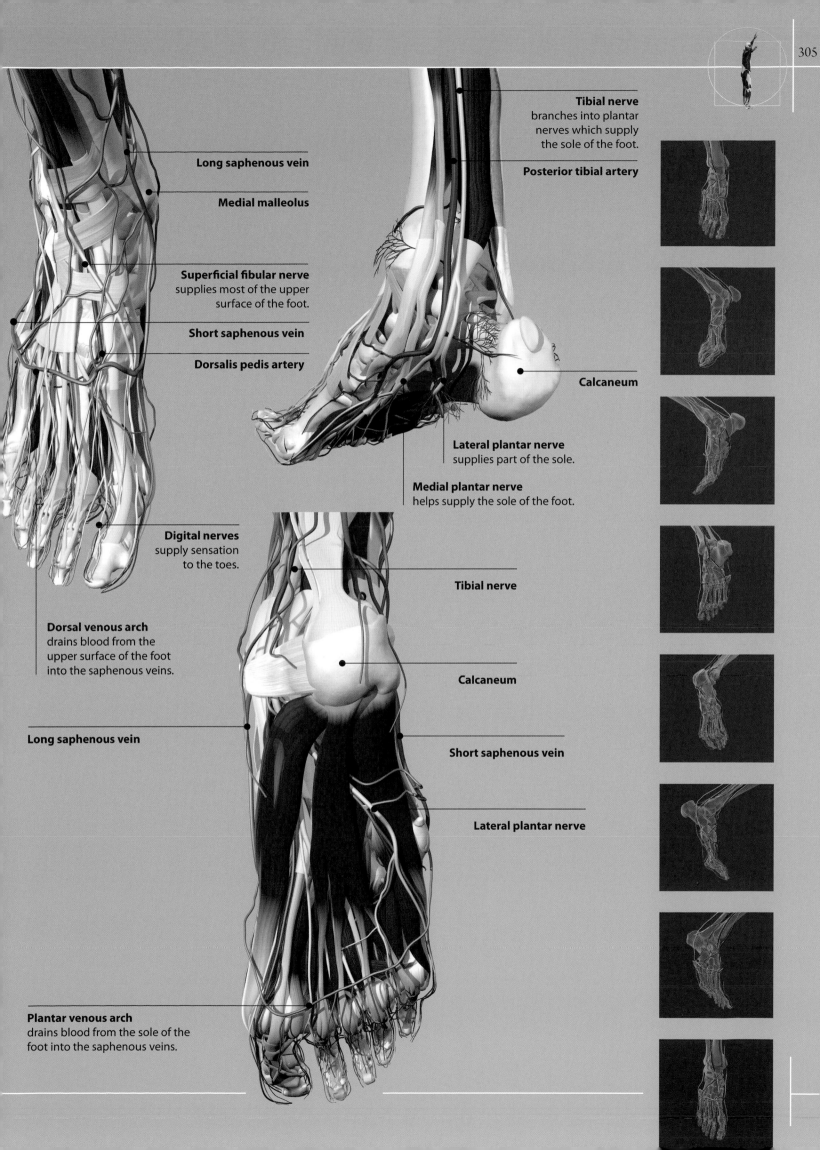

Long saphenous vein

Medial malleolus

Superficial fibular nerve supplies most of the upper surface of the foot.

Short saphenous vein

Dorsalis pedis artery

Tibial nerve branches into plantar nerves which supply the sole of the foot.

Posterior tibial artery

Calcaneum

Lateral plantar nerve supplies part of the sole.

Medial plantar nerve helps supply the sole of the foot.

Digital nerves supply sensation to the toes.

Dorsal venous arch drains blood from the upper surface of the foot into the saphenous veins.

Long saphenous vein

Tibial nerve

Calcaneum

Short saphenous vein

Lateral plantar nerve

Plantar venous arch drains blood from the sole of the foot into the saphenous veins.

The sole of the foot (plantar surface) distributes the weight of the body as we walk. Most of the small muscles of the foot are found in the sole where they are arranged into different layers. In addition, many blood vessels, nerves, and other structures pass through this region.

Plantar aponeurosis is a layer of tough fibrous tissue lying just under the skin. It protects the deeper structures of the foot, as well as helping to maintain the shape of the foot.

Abductor digiti minimi helps move the little toe out to the side.

Flexor digitorum brevis helps bend the toes.

Abductor hallucis helps spread the big toe away from the other toes.

Calcaneum

Tendons of flexor digitorum longus pass into the foot to help bend the toes.

Tendon of flexor hallucis longus helps bend the big toe.

Synovial sheaths surround and lubricate the long tendons of the leg muscles.

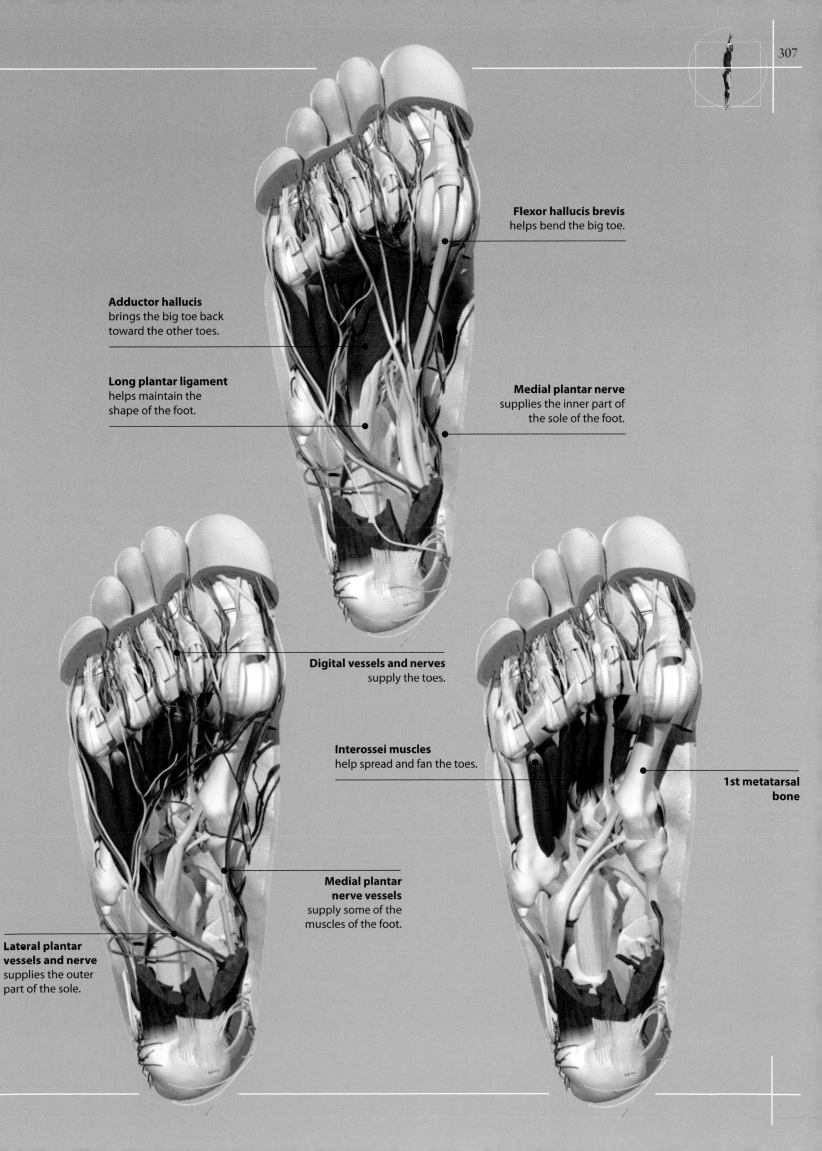

Flexor hallucis brevis
helps bend the big toe.

Adductor hallucis
brings the big toe back
toward the other toes.

Long plantar ligament
helps maintain the
shape of the foot.

Medial plantar nerve
supplies the inner part of
the sole of the foot.

Digital vessels and nerves
supply the toes.

Interossei muscles
help spread and fan the toes.

**1st metatarsal
bone**

**Medial plantar
nerve vessels**
supply some of the
muscles of the foot.

**Lateral plantar
vessels and nerve**
supplies the outer
part of the sole.

INDEX

B

D

F

G

H

I

J

L

K

M

N

O

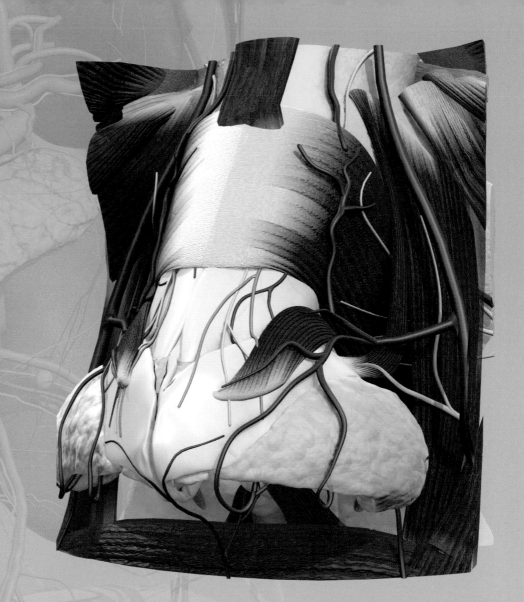

Q

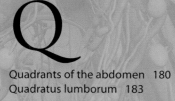

R

S

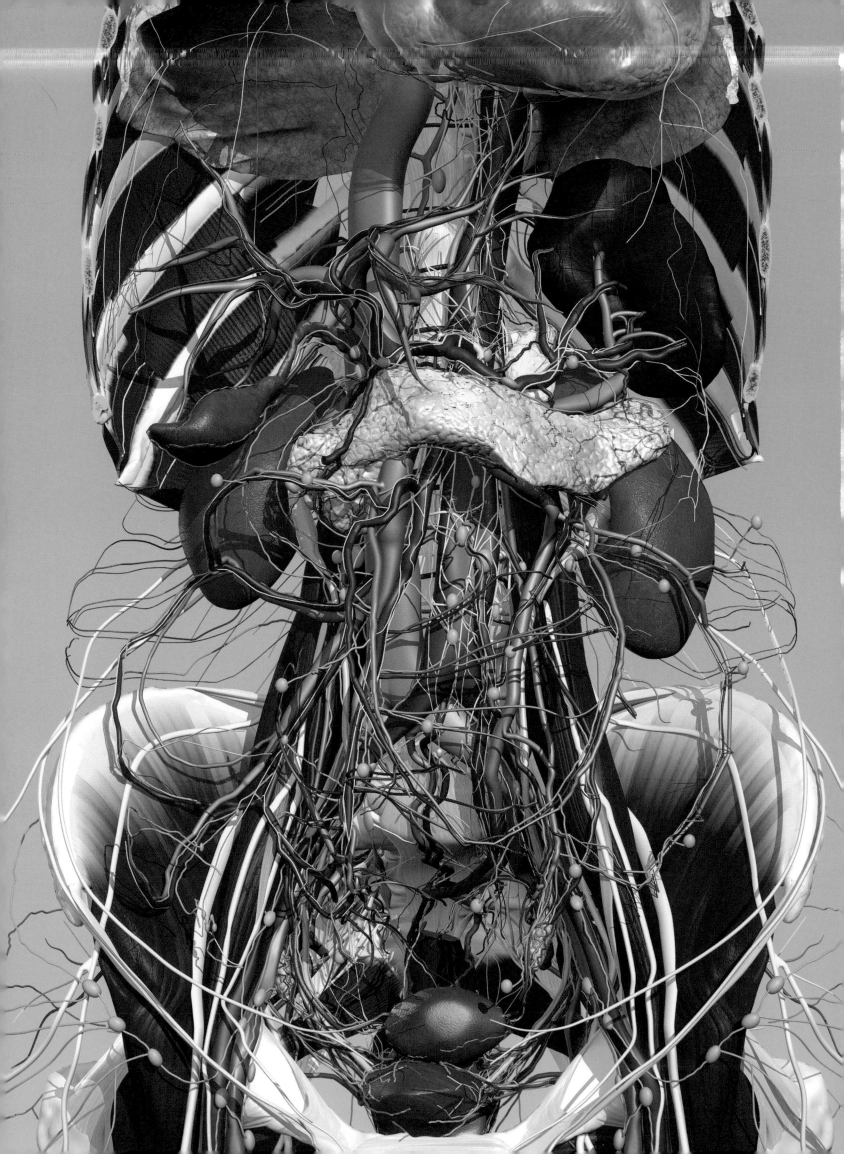